AF545089

ADVANCES IN PROTEIN CHEMISTRY AND STRUCTURAL BIOLOGY

Volume 81

Recent Advances in Electron Cryomicroscopy, Part A

ADVANCES IN PROTEIN CHEMISTRY AND STRUCTURAL BIOLOGY

Recent Advances in Electron Cryomicroscopy, Part A

EDITED BY

STEVEN J. LUDTKE

AND

B. V. VENKATARAM PRASAD

Verna and Marrs McLean Department of Biochemistry and Molecular Biology
Baylor College of Medicine
One Baylor Plaza
Houston, Texas 77030 -3498

ELSEVIER

AMSTERDAM • BOSTON • HEIDELBERG • LONDON
NEW YORK • OXFORD • PARIS • SAN DIEGO
SAN FRANCISCO • SINGAPORE • SYDNEY • TOKYO
Academic Press is an imprint of Elsevier

Academic Press is an imprint of Elsevier
The Boulevard, Langford Lane, Kidlington, Oxford OX5 1GB, UK
30 Corporate Drive, Suite 400, Burlington, MA 01803, USA
525 B Street, Suite 1900, San Diego, CA 92101-4495, USA

First edition 2010

ISBN: 978-0-12-381357-2
ISSN: 1876-1623

For information on all Academic Press publications
visit our website at www.elsevierdirect.com

Printed and bound in USA
10 11 12 10 9 8 7 6 5 4 3 2 1

Contents

From Envelopes to Atoms: The Remarkable Progress of Biological Electron Microscopy

R. Anthony Crowther

Present and Future of Membrane Protein Structure Determination by Electron Crystallography

Iban Ubarretxena-Belandia and David L. Stokes

Single-Particle Applications at Intermediate Resolution

Bettina Böttcher and Katharina Hipp

Visualizing Molecular Machines in Action: Single-Particle Analysis with Structural Variability

Sjors H. W. Scheres

FROM ENVELOPES TO ATOMS: THE REMARKABLE PROGRESS OF BIOLOGICAL ELECTRON MICROSCOPY

By R. ANTHONY CROWTHER

Medical Research Council Laboratory of Molecular Biology, Cambridge, United Kingdom

Abstract

The electron microscope has, in principle, provided a powerful method for investigating biological structures for quite sometime, but only recently is its full potential being realized. Technical advances in the microscopes themselves, in methods of specimen preparation, and in computer processing of the recorded micrographs have all been necessary to underpin progress. It is now possible with suitable unstained specimens of two-dimensional crystals, helical or tubular structures, and icosahedral viruses to achieve resolutions of 4 Å or better. For nonsymmetrical particles, sub-nanometer resolution is often possible. Tomography is enabling detailed pictures of subcellular organization to be produced. Thus, electron microscopy is now starting to rival X-ray crystallography in the resolution achievable but with the advantage of being applicable to a far wider range of biological specimens. With further improvements already under way, electron microscopy is set to be a centrally important technique for understanding biological structure and function at all levels—from atomic to cellular.

ADVANCES IN PROTEIN CHEMISTRY AND STRUCTURAL BIOLOGY, Vol. 81
DOI: 10.1016/S1876-1623(10)81001-6

I. Introduction

The development of powerful physical techniques for the determination of the structure of biological materials has almost always involved a long gestation period between the conception of the basic ideas and the realization of fully productive approaches. This was true for the development of macromolecular X-ray crystallography and NMR and is certainly true for high-resolution electron microscopy. In each case, many advances in basic instrumentation, in specimen preparation, and in computational analysis and interpretation of the experimental data were essential for the full potential of the technique to be realized. Developments in these different aspects can often proceed in parallel at different rates, but frequently, a breakthrough in one area stimulates a necessary advance in another area. Critical developments may often be conceptual, with full practical application coming much later and may, as so often happens, appear obvious in hindsight. Yet, each advance is a small triumph, giving pleasure to the inventor or discoverer and, taken in total, the small advances create a coherent and powerful approach to structure determination.

Here, I will give a personal view of the development of biological electron microscopy, highlighting what I see as some of the important advances. Inevitably, this will be a partial view, but I hope that any omissions or distortions will be corrected by the wide ranging and detailed accounts to be found in the succeeding chapters. I have been fortunate in my career to witness the flowering of the entire field of quantitative biological electron microscopy. For those who have entered the field more recently, it may be useful to recount some of the early history, as the full extent of the current success of electron cryo-microscopy can be better appreciated by reference to the more limited results of earlier years.

II. Early History

The story begins in Germany in the 1930s with invention by Ruska and colleagues of the electron microscope, an event recognized by the somewhat belated award to Ruska of the Nobel Prize in 1986. The early days are described in his Nobel lecture (Ruska, 1986) and it is notable that some of the first electron images of biological material were of bacteriophages. Subsequently, viruses, which are intrinsically interesting, readily purified, and possessed of various kinds of symmetries, have provided attractive specimens for many of the developments in imaging and analysis (Crowther, 2004). In turn, the electron microscope has revealed many important aspects of virus structure, some of which are described in this volume.

The problem with biological samples is that they are delicate, hydrated, and composed of atoms of low atomic number. It is therefore difficult to introduce them into the vacuum of the electron microscope; they are damaged by exposure to the electron beam and the images obtained are of low contrast. The first methods of contrast enhancement were based on shadowing with heavy metal atoms or positively staining with a heavy metal salt and washing away the excess salt. The dried specimens were robust and the images contrasty, but little was revealed apart from the particulate nature of the sample. Hall (1955) noted that better results could be obtained by omitting the washing step, thus allowing the particles to become surrounded by dense material. This was taken further by Huxley (1956), who visualized the central hole along the axis of tobacco mosaic virus, where the stain had entered, and noted that the 'outlining' technique would be useful for this type of specimen, particularly as it was so simple and gave excellent contrast and resolution. Brenner and Horne (1959) formalized the method and called it negative staining. A virus preparation was mixed with 1% phosphotungstate and sprayed onto a thin carbon film on the microscope grid and allowed to dry. The virus particles became embedded in a thin coat of stain, which for the first time revealed molecular details on the virus surface. Negative staining became a standard way of preparing particulate material and remains in use today as a simple and quick method of preparing robust specimens for electron microscopy. The fidelity with which the detailed shape of the surface of the particle is revealed is remarkable, but the definition of the internal molecular structure is extremely limited, as it is the stain rather than the biological material that gives the principal contribution to the image.

The advent of negative staining meant that the images were now sufficiently detailed to warrant a structural interpretation. Initial attempts at understanding the structure of viruses and their images were based on physical model building, using stick-like models to create a shadow image that mimicked the superposition of features in the projected view given by the electron image (Fig. 1; Klug and Finch, 1965). Around this time, computer-controlled film plotters were becoming available, so much

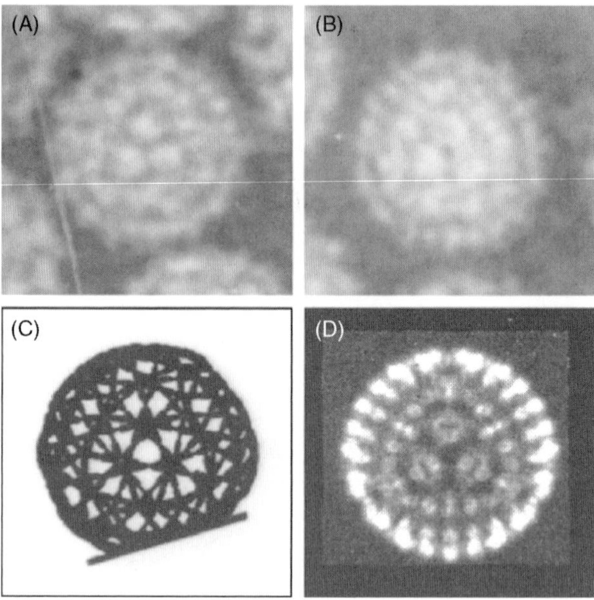

FIG. 1. Examples of model building to interpret images of human wart virus. (A, B) Images of negatively stained particles. (C) Shadowgraph (Klug and Finch, 1965) and (D) computer simulation (Klug and Finch, 1968). All views are down a threefold symmetry axis. Reprinted from the cited publications with permission from Elsevier.

more realistic projections could be created (Fig. 1; Klug and Finch, 1968). However, model building involved trial and error and there was no guarantee that any model could be invented to explain all of the features seen in the images. More direct and quantitative approaches were needed and their development had already started.

The recorded image of the biological specimen is degraded by extraneous noise arising from the supporting carbon film, from the granularity of the stain and, in low dose images used to minimize the radiation damage, from statistical fluctuations in the number of electrons in each image element. If the specimen is made from repeated units arranged in a symmetrical way, as is often the case for macromolecular assemblies, it is possible to enhance the signal and reduce the noise by averaging over the repeated copies of the unit in the image. The first attempt at doing this was made by Markham et al. (1963) using photographic superposition for rotationally symmetric images, and in the following year, they described a photographic linear integrator for averaging images with translational periodicity (Markham et al. 1964). In each case, the method both determined the periodicity of the dominantly repeating features in the image, rotational or translational, and simultaneously created an enhanced image. The problem was that the appropriate symmetry for averaging had to be determined by trial and error, and the judgement of what was significant was made subjectively by looking at the differently averaged images, which could be misleading.

What was needed was a more objective method, in which the analysis of symmetry was separated from the creation of an averaged image. Klug and Berger (1964) took the first step by using an optical diffractometer, a device introduced by Lipson and Taylor for the interpretation by simulation of the X-ray diffraction patterns of crystals, to generate an optical diffraction pattern or Fourier transform of the micrograph. Recording the diffraction pattern captures the strengths of all the Fourier components in the image and allows any dominant translational periodicities to be detected. This was the first time that Fourier transforms had been used to analyze micrographs, and the development proved to have great utility. Introduction of a filter mask in the diffraction plane and recombination of those diffracted rays allowed through the mask created a filtered image (Klug and DeRosier, 1966). The size of holes in the mask controlled the range of averaging in the image, with smaller holes giving a greater degree of averaging. For specimens consisting of two layers, such as would be formed by a collapsed tubular structure, the two layers could be separated in the filtered image by allowing through the mask just those diffracted beams corresponding to one of the layers. The two sides of a helical structure could also be separated in the same way.

In a later development for averaging rotationally symmetric structures, the separation of the steps of analysis of symmetry and synthesis of an averaged image was carried out computationally by decomposition into and synthesis from a set of angular harmonics (Crowther and Amos, 1971). The strengths of the different harmonics could be plotted as a rotational power spectrum, with the strongest peaks showing the dominant symmetry. This was analogous to the peaks in the optical diffraction pattern, which showed the dominant translational symmetry. In each case, the procedure was more quantitative and less subjective than the Markham type of photographic superposition.

III. Three-Dimensional Reconstruction

Even after the sides of a helical structure, such as the tail of bacteriophage T4, had been separated by filtering, it was clear that the filtered image still exhibited a substantial overlap of features at different cylindrical radii. It was at this point that the key advance was made and the whole field of quantitative computer image processing was initiated. It is not often that the start of a field can be so precisely ascribed, but in this case, the paper by DeRosier and Klug (1968) clearly marks the beginning of developments that have eventually led to the results described in the present volume. Their paper was entitled "Reconstruction of three-dimensional structures from electron micrographs", and although the method was applied to the special case of a helically symmetrical specimen, the tail of bacteriophage T4, the general applicability of the approach was emphasized. They wrote,

> "Our method starts from the obvious premise that more than one view is generally needed to see an object in three dimensions. We determine first the number of views required for reconstructing an object to a given degree of resolution and find a systematic way of obtaining these views. The electron microscope images corresponding to these different views are then combined mathematically, by a procedure which is both quantitative and free from arbitrary assumptions, to give the three dimensional structure in a tangible and permanent form. The method is most powerful for objects containing symmetrically arranged subunits, for here a single image effectively contains many different views of the structure. The symmetry of such an object can be introduced into the process of reconstruction, allowing the three dimensional structure to be reconstructed from a single view, or a small number of views. In principle, however, the method is applicable to any kind of structure, including individual unsymmetrical particles, or sections of biological specimens."

The key points in the paper were the recognition that the electron micrograph represented a projection of scattering material in the direction of the electron beam; that the projection data could be conveniently combined as central sections of the three-dimensional Fourier transform; that the number of different views necessary to make the three-dimensional map could be determined depending on the size of the object and the resolution desired in the final map; and that by computing a complex numerical Fourier transform of the digitized micrograph, both amplitude and phase information could be recovered from the image. There was thus no "phase problem" of the kind that confronted X-ray crystallographers, where only diffracted intensities could be measured and phases had to be recovered indirectly by use of heavy atom derivatives.

It was at this time that I became peripherally involved with these developments. In order to be able to process micrographs computationally, it was necessary first to convert the image into an array of numbers representing the optical density. As a graduate student, I had been writing programs to control a flying spot densitometer so that it could be used to measure the intensity of spots on X-ray crystallographic diffraction patterns (Arndt et al., 1968). At that time, the programs for instrument control were all written in machine code for a Ferranti Argus computer. Nevertheless, it was easy for me to provide a program to scan selected areas of micrographs, as needed for the image processing.

The approach now was wholly computational. Computers were, by this time, fairly widely used in structural biology for computing electron density maps from X-ray crystallographic data, and it is no accident that three-dimensional reconstruction was invented in the laboratory that had earlier seen pioneering work in the development of protein crystallography. The core of the reconstruction method depended on a relationship well known to crystallographers, namely that the two-dimensional Fourier transform of a projection of a three-dimensional structure corresponds to the equivalent central section through the three-dimensional transform. Thus, the different views give different central sections, so with sufficient views, the three-dimensional transform could be filled in completely and the density in the object recovered by Fourier synthesis. The case of a helical structure was special because the specimen effectively contained a tilt axis, so that a single view presented a set of equally spaced views of the repeating subunit. The two-dimensional transform of a single view thus contained sufficient information to make a three-dimensional map, at least to limited resolution. The data were analyzed using Fourier–Bessel theory already developed for patterns from X-ray fiber diffraction of helical specimens (Klug et al., 1958). The three-dimensional map was actually calculated using the program written for making a map of TMV from X-ray diffraction data. This underlines again the close interplay at that time between X-ray methods and electron microscopy. Now that maps from electron micrographs are reaching atomic resolution, it is already becoming profitable to reestablish the connection with X-ray crystallography for map display, model building, and structure refinement. The original map of the T4 tail (Fig. 2) was constructed of balsa wood and

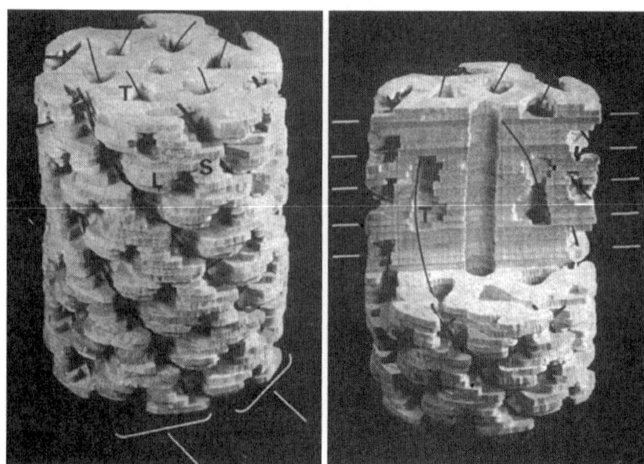

Fig. 2. Three-dimensional map of the T4 phage tail (DeRosier and Klug, 1968). The density in the map is represented by a set of glued balsa wood sections. Reprinted by permission from Macmillan Publishers Ltd: *Nature*, DeRosier and Klug, copyright 1968.

the model is housed in the Science Museum in London, as befits the first example of a completely new approach to structure determination.

For nonhelical objects, it was necessary to combine more than one image of the specimen. These images could come from particles lying in different orientations on the grid or be collected from a single particle by tilting in the microscope. In either case, any internal symmetry in the particle helps to reduce the number of different views required and also helps with other aspects of the computer processing. Accordingly, the next development, in which I was closely involved, was to make maps of icosahedral viruses. With icosahedral symmetry, one general view of a particle gives rise to 60 symmetry-related planes in the three-dimensional transform. However, compared with the case of helical symmetry, the data points are very unevenly distributed, a problem that is exacerbated by the inclusion of data from multiple views in arbitrary orientations. We therefore had to develop methods for interpolating and combining such unevenly sampled data to create a representation of the three-dimensional transform that could be properly inverted to make a density map (Crowther et al., 1970b). The methods we proposed depended on the finite size of the object, which limits how fast the transform can vary and gives rise to interpolation formulae of the Whittacker–Shannon type.

In fact, for spherical viruses, we reverted to the kind of analysis used for helical structures and represented the transform by cylindrical harmonics, expressing only the 522 sub-symmetry of the full icosahedral 532 symmetry (Crowther, 1971). This led to a simpler computation for the interpolation, important given the limited computing power then available, and meant that we could use a modified version of the helical Fourier program for computing the final map.

Maps were computed of tomato bushy stunt virus (Fig. 3) and human wart virus (Crowther et al., 1970a). These showed clearly the arrangement of morphological units, although in the latter case, the resolution achieved was not sufficient to show, as later emerged, that all the capsomeres were pentamers (for a later map, see Fig. 9), not the pentamers and hexamers expected on the theory of Caspar and Klug (1962). We were puzzled at the time that the 5-coordinated units were of the same size as that of the 6-coordinated units. These papers also introduced the idea of "common lines", which arise in the two-dimensional transform of the image of a symmetrical structure from the intersection of symmetry-related planes in the three-dimensional transform. These can be used to find the orientation and center of any view of the virus relative to the symmetry axes, parameters that are essential to determine before the data can be combined. Cross-common lines between different views can be used for interparticle scaling and for ensuring that the different views are combined with a consistent choice of hand. The absolute hand has to be determined by tilting experiments. Some of these basic ideas, although

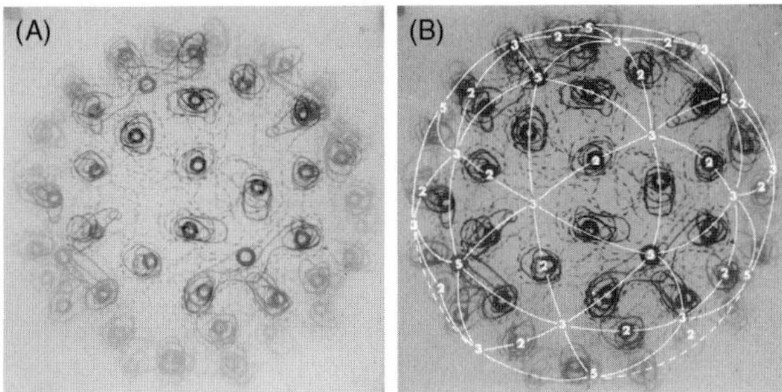

Fig. 3. Map of tomato bushy stunt virus (Crowther et al., 1970a,b), in which the 180 protein subunits are clustered in 90 dimers. In (A), the dimeric units are indicated and in (B), the icosahedral symmetry axes are marked.

introduced specifically to apply to icosahedral viruses, have proved to be more generally applicable for particles of all kinds. Despite giving useful information about various macromolecular assemblies, the amount that could be learned was limited by the staining technique itself to molecular envelopes at a level of detail of about 15–20 Å. A good snapshot of the state of the field at this time is provided by the published proceedings of a conference entitled "New Developments in Electron Microscopy" (Huxley and Klug, 1971). However, a better way of preparing specimens for microscopy was badly needed.

IV. Unstained Crystals

The challenge was to find a way of replacing water that preserved the biological structure and also to choose specimens that were amenable to the kinds of imaging and data processing that would be needed. Glucose embedding provided a simple means of specimen preparation that preserved the structure and enabled the specimen to be placed in the vacuum of the microscope (Unwin and Henderson, 1975). Two-dimensional crystals provided an appropriate specimen. Previous studies (Taylor and Glaeser, 1974) had shown that electron diffraction patterns showing spots corresponding to 3.4 Å spacings could be obtained from frozen hydrated protein crystals, indicating that a high degree of order could be preserved in the microscope, but the specimens were tricky to deal with and no images were collected. With glucose embedding, it became possible using a normal microscope at room temperature to collect high-resolution diffraction patterns and images from unstained specimens of catalase and purple membrane (Unwin and Henderson, 1975). In the absence of a protective coat of stain, the specimens were highly radiation sensitive, so they had to use low dose techniques. Their paper gives a lot of intimate details about how to set up and use the microscope to achieve this end.

With the low doses used ($\sim$1 e/Å^2), the image was dominated by the statistical noise from the limited number of electrons per image element. No structure was detectable by eye and the micrographs looked essentially like neutral density filters. However, because the specimens were crystalline, when placed in an optical diffractometer, the images gave a set of sharp diffraction spots. This indicated good preservation of the crystalline order, so that a reliable image of the repeating unit could be generated by averaging over many unit cells. Not only was the specimen highly radiation sensitive, it also gave images with very low contrast, because all the scattering was now from light atoms. This meant that it was necessary to introduce phase contrast by recording images with the microscope in an underfocussed state. Because underfocus creates a rather imperfect form of phase contrast, in which alternating bands of spatial frequencies contribute with opposite signs of contrast, extensive image processing was needed to extract the high-resolution information. This kind of approach had already been used by Erickson and Klug (1971) to compensate for underfocus in images of negatively stained catalase crystals, but in that case, the improvement achieved was limited by the stained nature of the specimens. In the case of the glucose embedded samples, far more detail was apparent in the averaged projections, which now related directly to protein and not stain and which were determined to 7 Å for purple membrane and 9 Å for catalase (Unwin and Henderson, 1975). The projection structure of the purple membrane turned out to be particularly informative, as it indicated that a large part of the constituent bacteriorhodopsin molecule consisted of seven α-helices running roughly normal to the plane of the membrane.

This interpretation was confirmed soon afterward by production of a three-dimensional map (Fig. 4) at a resolution of 7 Å (Henderson and

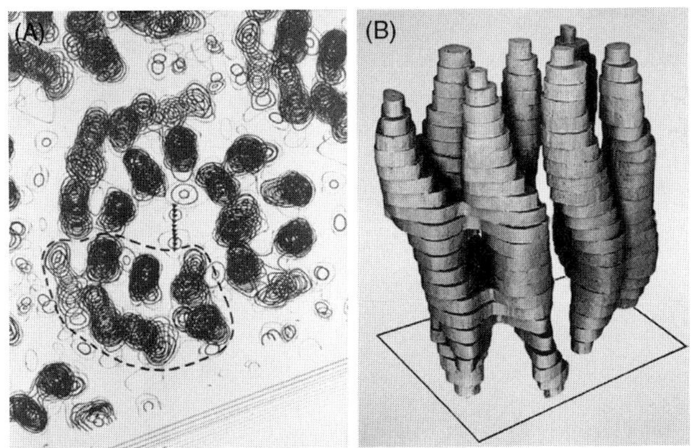

Fig. 4. Map and model of bacteriorhodopsin (Henderson and Unwin, 1975). (A) Part of the three-dimensional map showing a region 50 Å thick spanning the membrane, with one of the bacteriorhodopsin molecules outlined. (B) Model showing the seven transmembrane helical segments. Reprinted by permission from Macmillan Publishers Ltd: *Nature*, Henderson and Unwin, copyright 1975.

Unwin, 1975). Although the loop connectivity of the various helices was not clear, this was for the first time that the internal structure of a protein had been determined by electron microscopy and the first structure for a membrane protein, so it represents a landmark in the development of the subject (Fig. 4). Further technical developments, such as the use of a field emission gun (FEG), spot-scan recording of the images, correction of the defocus gradient across the image of a tilted specimen, unbending of distorted crystal lattices, and dealing appropriately with electron optical defects such as astigmatism and beam tilt, eventually led to an atomic model for the structure (Henderson et al, 1990). It had taken 15 years to move from tubes of density representing α-helices to a map that showed amino acid side chains, indicating the number of technical problems that had to be solved. The data were sufficiently accurate and overdetermined that the structure could be subjected to crystallographic refinement (Grigorieff et al., 1996), to produce a model whose details rivalled those obtained by X-ray protein crystallography (Fig. 5). Serendipitously, what might at the outset have been just an obscure bacterial proton pump turned out to be a paradigm for an extremely important class of eukaryotic membrane proteins, the G-protein coupled receptors. The power of electron diffraction for analyzing well-ordered two-dimensional crystals was demonstrated by the 1.9 Å refined map of aquaporin, which showed details of the pore and of interactions between protein and lipids (Gonen et al., 2005). These examples fully realize earlier suggestions about the possibility of using the electron microscope for crystallographic structure analysis (Hoppe et al., 1968; Hoppe, 1970).

V. Rapid Freezing

Glucose embedding provided a highly effective way of preparing two-dimensional crystalline specimens, in which a large area of crystal was required to lie flat on the carbon support film. It was much less effective for particulate specimens such as viruses because of low-resolution contrast matching between glucose syrup and protein. An alternative approach was needed and it was provided by Dubochet and colleagues at the European Molecular Biology Laboratory (for review, see Dubochet et al., 1988). What they did was to place a small drop of the particle suspension on a holey carbon film, blot off most of the liquid to leave a thin film, and then quickly plunge the grid into ethane

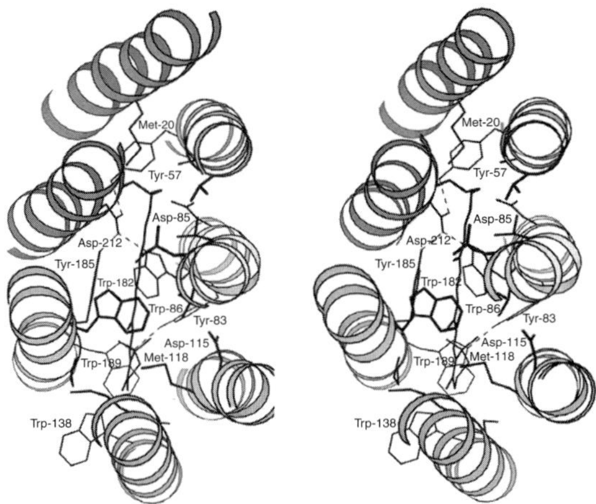

FIG. 5. Stereo view of the retinal binding pocket in bacteriorhodopsin (Grigorieff et al., 1996). This shows details of the amino acid side chains involved in retinal binding and thus in the chemistry of light adsorption and proton pumping. Reprinted from the cited publication with permission from Elsevier.

slush held at liquid nitrogen temperature. The film of liquid is so thin (~500–1000 Å) that it freezes very rapidly and turns into solid water without crystallizing. This glass-like amorphous or vitreous ice provides a very good hydrated environment to preserve the details of the biological structure. The specimen once frozen must be maintained at liquid nitrogen temperature during loading into the microscope and during data collection, so as to avoid crystallization of the water, and care must be taken to avoid contamination on the surface of the ice. However, if the technical difficulties can be overcome, this kind of specimen does give the opportunity for investigating biological mechanisms in a physiologically relevant context.

Because the scattering density of the amorphous ice is lower than glucose, the particles stand out from the background. The vitreous ice itself has a very smooth appearance in images and by using particles suspended in ice over holes in the carbon film there is little extraneous contribution to the image of the particle. Among the earliest particulate specimens examined by Dubochet and colleagues was a range of viruses and spectacular images of native structures were obtained (Adrian et al., 1984). Because the contrast is still weak, imaging requires underfocus to produce phase contrast. Of course, the specimens are very radiation sensitive, necessitating low dose recording of images. These are the same basic problems that had to be overcome with glucose embedding but now one does not have the extended translational repeat of the crystal to help with the averaging. Nevertheless with symmetrical structures, such as helical tubes and icosahedral viruses, considerable progress has been made toward achieving high resolution, as will be described next. The case of particles with low or no symmetry, which requires rather different methods of analysis, will be addressed later. For all kinds of specimens, the development of FEG microscopes has been a key advance in providing images with much improved transfer of high-resolution information from sample to micrograph. Much of the detailed background to the collection and processing of images of cryo-specimens is covered in the monograph by Glaeser et al. (2007).

VI. Helical Structures

Specimens with helical symmetry offer many of the advantages of two-dimensional crystals in that a single view provides multiple views of the constituent subunit, related in a symmetrical way, as was noted originally (DeRosier and Klug, 1968). However, because of distortions in the extended structure, the degree of detail obtained by straightforwardly applying the helical symmetry throughout the processing may be limited. To overcome this, it may be profitable to split the helical structure into a series of adjoining or overlapping segments, which can then be processed locally as single particles (Beroukhim and Unwin, 1997; Egelman, 2000), while still applying some of the helical constraints (Sachse et al., 2007). By fitting to a reference structure, the in-plane rotation, the out-of-plane tilt, the variation in twist, the displacements of the origin, and the radial scaling can be determined for each segment, provided the signal-to-noise ratio is adequate. The segments must therefore be sufficiently large for an accurate determination of these parameters, while not being so big that the segments themselves contain significant distortions.

Using this approach, the structure of the acetylcholine receptor embedded in membrane tubes was eventually determined to 4 Å (Miyazawa et al., 2003) and then refined (Unwin, 2005). The atomic model gave a detailed description of the whole receptor in its closed configuration (Fig. 6), including the intracellular and ligand binding domains, which had not previously been interpreted in detail. Using helical image analysis of the

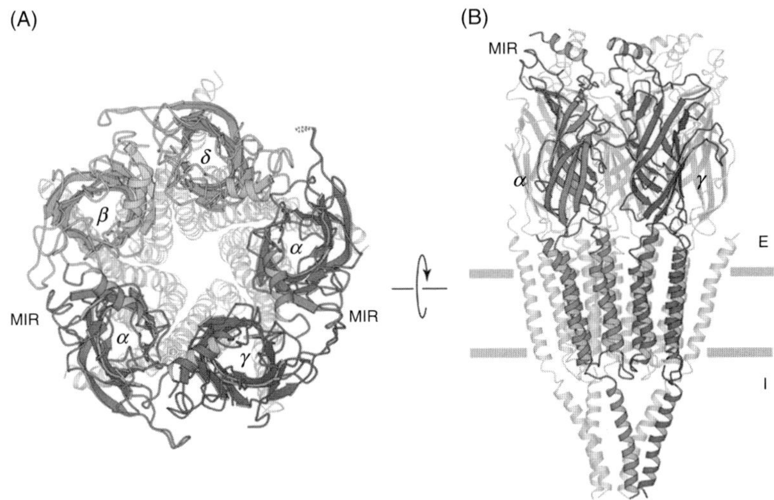

FIG. 6. Molecular model of the acetylcholine receptor (Unwin, 2005). (A) View down the pseudo-fivefold axis of the receptor, normal to the plane of the membrane and (B) view normal to the pseudo-fivefold axis. The $\alpha_2\beta\gamma\delta$ subunits are indicated and in (B), the interior (I) and exterior (E) of the membrane are marked. Reprinted from the cited publication with permission from Elsevier.

bacterial flagellar filament, an atomic model for the structure of one conformation of the filament was produced (Yonekura et al., 2003). The structure of tobacco mosaic virus was "revisited" (Sachse et al., 2007) and a map produced at about 4.5 Å. Compared with previous maps from X-ray diffraction from fibers of virus or from crystals of protein disks, the new map showed improved detail at inner radii and highlighted structural elements that might be important in assembly and disassembly. A particularly interesting example, though not yet at such high resolution, is provided by a bacterial dynamin-like protein assembled on a lipid tube, in which large conformational changes compared with a crystal structure of the isolated molecule have important implications for function (Low et al., 2009). This example illustrates the power of combining atomic maps from crystallography with lower resolution maps from cryo-EM to gain important functional information about states that might be otherwise inaccessible to structural investigation, a topic that will be examined in more detail in Chapter 3. Recent advances in determining helical structures are covered in Chapter 2.

VII. Amyloids

In various human diseases, normally soluble proteins or fragments of proteins assemble into filaments termed amyloids. Such abnormal assemblies are believed to interfere with cellular functions and in many cases to lead to the symptoms of the disease. This is particularly true in the case of various common neurodegenerative diseases. In Parkinson's disease and associated movement disorders, the normally soluble but natively unfolded protein α-synuclein assembles into filaments that form large intracellular aggregates termed Lewy bodies. In Alzheimer's disease and allied dementias, it is the microtubule associated protein tau that assembles into filaments that form intracellular neurofibrillary tangles. Also in Alzheimer's disease, a proteolytic fragment of a large membrane protein, the amyloid precursor protein (APP), aggregates into so-called β-amyloid that forms extracellular deposits termed plaques. Most cases of these neurodegenerative diseases are sporadic, but for α-synuclein, tau, and APP, there are rare dominantly inherited mutations in the respective genes that give rise to disease, indicating the importance of these proteins in the aetiology of the respective conditions.

Electron microscopy has proved a valuable tool for investigating filament formation and structure. Most amyloid filaments exhibit a twisted morphology but the repeat or cross-over spacing can be quite variable, even along a given filament, and the surface of the filament generally appears smooth. The underlying architecture of amyloids is based on a cross-β fold, in which the protein chain runs approximately at right angles to the filament axis and the β-strands then stack to form β-sheets. Once formed, the filaments are very stable. X-ray or electron diffraction patterns show a characteristic spacing at 4.7 Å in an axial direction, arising from the separation of cross-β strands, and may show equatorial spacings in the range 10–12 Å, arising from the stacking of β-sheets. Selected area electron diffraction is particularly useful as only a small area of specimen is used, and this can afterward be imaged to view the diffracting species and establish the relationship between the structure and the diffraction pattern. Diffraction from a small raft of *in vitro* assembled and partially aligned α-synuclein filaments embedded in ice is shown in Fig. 7 (Serpell et al., 2000). The diffraction pattern shows strong arcs corresponding to spacings of 4.7 Å and a second order of 2.35 Å in the axial direction of the filaments. In glucose-embedded filaments, the third order of the cross-β spacing at 1.15 Å can sometimes be seen. Taken together with the sharpness of the reflections, this indicates the high degree of axial order in the cross-β structure of the filaments. Tau filaments could be treated similarly (Fig. 7), though in this case, native filaments from human brain were used (Berriman et al., 2003). The image formed in defocused diffraction mode indicates that oriented diffraction arcs at 4.7 Å can be recorded from just a few filaments, whose morphology can be discerned. In the most extreme case, diffraction could be recorded from what appeared to be just a single filament. With the rather crude material from Alzheimer brain, as opposed to *in vitro* assembled filaments, it was important to be sure that the diffraction was coming from tau filaments and not from β-amyloid. To be able to measure diffraction from such a small sample and moreover to be able to relate the

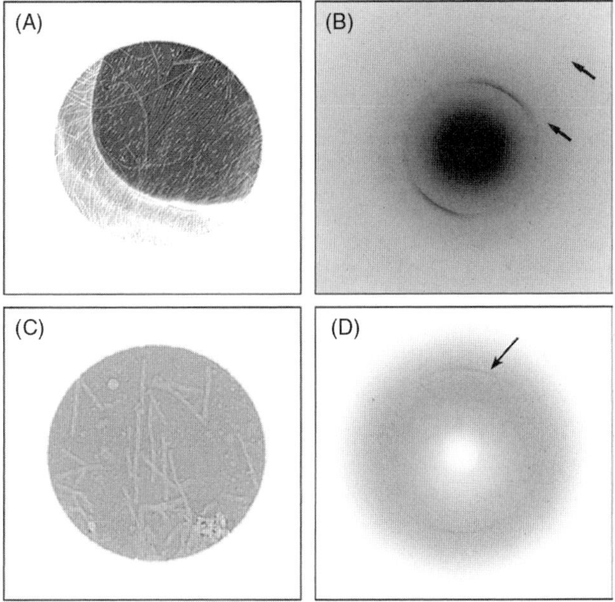

Fig. 7. Selected area electron diffraction from amyloid filaments. (A) Raft of *in vitro* assembled α-synuclein filaments and (B) diffraction pattern from this area (Serpell et al., 2000). The arrows indicate arcs corresponding to axial spacings of 4.7 and 2.35 Å in the cross-β structure of the filament. (C) Tau filaments from Alzheimer brain and (D) corresponding diffraction pattern with the 4.7 Å spacing arrowed (Berriman et al., 2003).

diffraction pattern to the morphology of the specimen are clearly a huge advantage compared with X-ray diffraction from a bulk sample which cannot be imaged.

Because of the variability in twist in amyloid filaments, it has been difficult to compute three-dimensional maps. There is also a problem that images of the filaments show very little structural detail between the long range twist and the short range spacings of the cross-β fold. Cross-sections computed from negatively stained specimens of two kinds of tau filaments from Alzheimer brain indicated that the two morphologies arose from different packings of a common subunit (Crowther, 1991). Higher resolution images of amyloid filaments had to await a more sophisticated analysis of unstained specimens of β-amyloid (Fig. 8; Sachse et al., 2008). Using single-particle alignment of small segments of filament, a map was computed to about 8 Å resolution. The cross-sections showed that the filament consisted of two protofilaments, each containing 50 Å long features that appeared to represent β-sheet structure (Fig. 8). Within each protofilament, a local twofold symmetry suggested that pairs of β-sheets were formed from equivalent parts of two Aβ peptides. Atomic structures from X-ray crystallography are available only for very short peptide spines (Sawaya et al., 2007), and it seems unlikely that long peptides or whole proteins will ever form amyloid crystals suitable for X-ray structure determination. Thus, to understand the pathologically relevant forms of filament, it will be essential to go on improving the maps from electron microscopy, incorporating information from other structural studies as appropriate.

VIII. Icosahedral Viruses

Icosahedral viruses provided a good test specimen for the early developments of image processing, and with the advent of cryo-imaging, the challenge was to improve the processing to take advantage of the potentially higher resolution information derivable from the images. The overall scheme for processing images of icosahedral viruses is shown in Fig. 9. Initially, after a particle has been selected, its approximate centre and orientation relative to the symmetry axes can be found by self-common lines. Because of the high noise level, in many cases, these parameters would not have been correctly determined. Nevertheless, an initial three-dimensional map can be calculated from the best subset of particles.

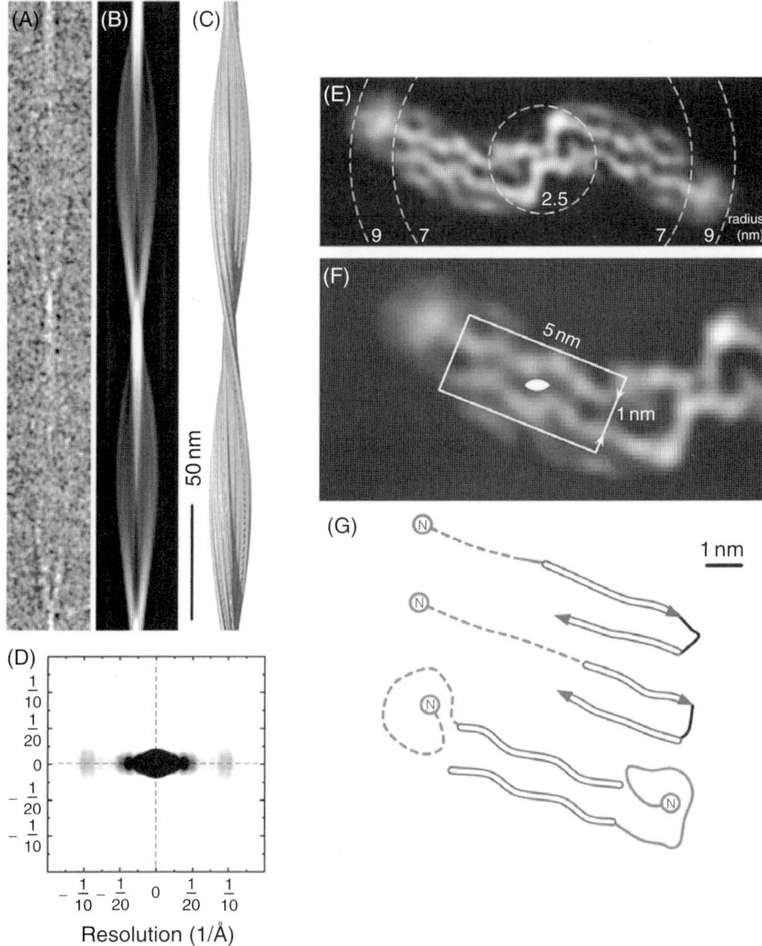

Fig. 8. Structure of β-amyloid fibril (Sachse et al., 2008). (A) Ice-embedded Aβ (1–40) fibril. (B) Projection of the three-dimensional reconstruction. (C) Surface rendering of the three-dimensional map. (D) Calculated electron diffraction from the three-dimensional map, showing equatorial reflections peaking at about 10 Å, corresponding to β-sheet spacing. (E) Fibril density projected along the helical axis, showing a twofold symmetry corresponding to two protofilaments forming fibril. (F) β-Sheet sandwich within one protofilament (boxed rectangle). (G) Model of possible Aβ (1–40) peptide packing within one protofilament. Reprinted with permission, copyright 2008 National Academy of Sciences, USA.

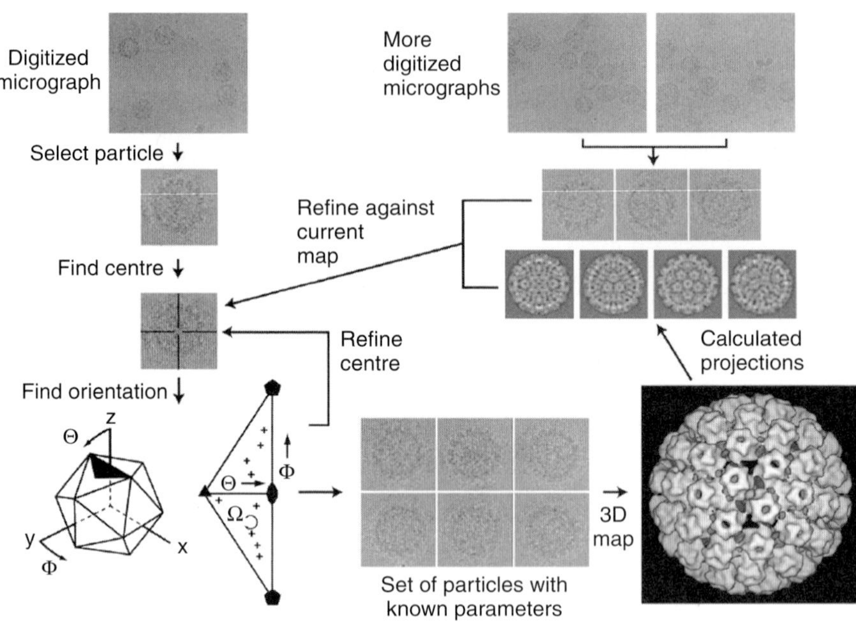

FIG. 9. Outline of scheme for processing icosahedral virus particles, with human wart virus as the example (Crowther, 2008). A preliminary map is calculated from an initial set of particle images, and the set of calculated projections forms a basis for finding origins and orientations of further particles by cross-common lines (see text). From these, an improved map is calculated and the whole process is iterated to achieve higher resolution.

Assuming that most of the parameters were correct, this preliminary map will be an approximate representation of the structure but with a much reduced level of noise compared with the original images, because data have been combined from different images and icosahedrally symmetrized. The computed projections of this map can then be used for model-based refinement of the origins and orientations of extra particles selected from additional micrographs. The model-based refinement can be done in Fourier space using cross-common lines between each raw image and a small set of projections (Crowther et al., 1994). Typically, six well-spaced projections of the model generate 360 cross-common lines in the two-dimensional transform of the raw image, so giving an average 1°

angular sampling of that transform, which should prove more than adequate. Alternatively, a whole asymmetric unit of projections can be calculated from the model and then two-dimensional projection matching can be carried out (Cheng et al., 1994; Cheng and Baker, 1996). Because of the reduced noise in the model and the much greater sampling of the raw data compared with using self-common lines, the model-based approach generates much more reliable view parameters for each particle, enabling a better map to be calculated. The improved map represents a better model for the next round of refinement and the whole process can then be used iteratively (Fig. 9) to work toward a higher resolution. The number of particles needed to get to high resolution now depends not simply on adequate geometrical sampling of the three-dimensional transform, as was largely the case with stained samples, but much more on the need to overcome the decreasing signal-to-noise ratio with increasing resolution in the cryo-images. At least with an unstained specimen, there is some hope of determining the details of the internal structure.

Hepatitis B virus is a major human pathogen that replicates in the liver. There are estimated to be 350 million carriers worldwide and it causes about 1 million deaths a year from cirrhosis and primary liver cancer. Hepatitis B virus is an enveloped particle with an icosahedral core containing the genome, in this case, partially double-stranded DNA. The envelope consists of a lipid membrane containing virally coded surface proteins. The core protein when expressed in bacteria assembles into particles of two sizes, containing either 180 ($T=3$) or 240 ($T=4$) subunits clustered as dimers that create 90 or 120 spikes on the surface of the core (Crowther et al., 1994). Using the methods outlined above, a map at 7.4 Å resolution was computed of the $T=4$ core shell (Fig. 10; Böttcher et al., 1997). The fold of the protein, which was largely α-helical, was deduced, and using biochemical and immunological data, an amino acid numbering scheme was proposed (Fig. 10). This was for the first time that this level of detail had been achieved by single-particle methods and for the first time, a protein fold had been solved in this way. As with myoglobin, the first X-ray crystal structure, and bacteriorhodopsin, the first two-dimensional EM crystal structure, the fold of the core protein being largely α-helical was hugely helpful in interpreting the map. The interpretation of the map was confirmed subsequently by the determination of the X-ray crystal structure of the core shell (Wynne et al., 1999).

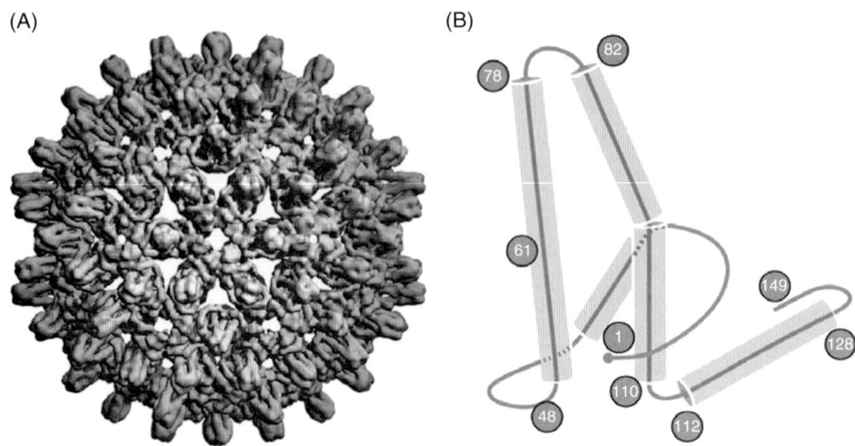

FIG. 10. Hepatitis B virus core protein shell (Böttcher et al., 1997). (A) Map at 7.4 Å resolution. The tubular features constituting the spikes on the surface are α-helices. (B) The numbered fold of the protein subunit deduced from the map, with cylinders representing α-helical regions. The complete shell is made up from 240 such subunits, paired to form 120 spikes.

The spikes on the surface of the core are formed by a bundle of four α-helices, the two apposed helical hairpins being contributed by the two molecules forming the dimer. This interpretation of the structure of the spike was also proposed by Conway et al. (1997), and remarkably, the Conway et al. and Böttcher et al. manuscripts arrived completely independently on the journal editor's desk within 24 h of each other, one according to its title using cold electrons, the other a cold microscope! This dichotomy of usage still prevails, perpetuated by the convenient form of cryo-EM as an abbreviation.

Hepatitis B virus is a pararetrovirus, so at various stages of its life cycle, its genome may be in the form of RNA or DNA, and reverse transcription is an essential part of that cycle. After the virus enters a hepatocyte, the partially double-stranded DNA in the core passes into the nucleus and various messenger RNAs are transcribed. When the viral polymerase/ reverse transcriptase is translated from a full length genomic RNA, it stays associated with this copy of the message. The separately synthesized core proteins assemble around this complex to form an immature core. The viral polymerase then copies the RNA into the first strand of DNA, degrading the RNA as it goes. Second-strand DNA synthesis takes place and by the time one-half to three-quarters of the second strand has been made the core becomes mature. At this stage, it can interact with surface proteins, which have been separately synthesized and inserted into an inner cellular membrane. The core buds through this assembly and becomes enveloped. Budding does not occur before second-strand DNA synthesis, so a maturation signal must be transmitted from the interior of the core, where DNA synthesis takes place, to the surface of the core, where interaction with the envelope takes place.

By comparing RNA containing cores made in bacteria, presumed to mimic an immature stage of the core, with mature DNA containing cores extracted from virions, a possible mechanism was identified (Roseman et al., 2005). Besides clear differences in the shape of the spike between the two structures, there are also small changes in other places in the shell. The shell acts as a mechanism by which changes on the inside are propagated through the protein to cause changes in the spike. One of the most significant changes is in a hydrophobic pocket that opens on the side of the spike (Fig. 11). The functional importance of this region is highlighted by the clustering there of various core mutations that affect viral secretion. Compensating mutations in viral surface proteins indicate that this is a probable site of interaction between core and surface proteins during assembly.

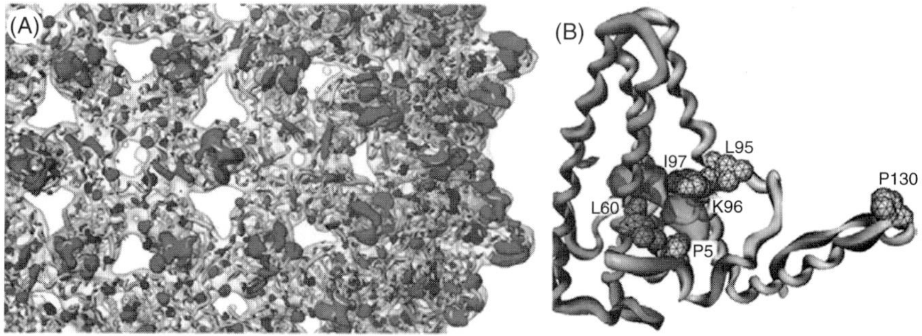

FIG. 11. Differences between maps of RNA and DNA containing hepatitis B core particles (Roseman et al., 2005). (A) Strongest significant differences superimposed on the RNA map, showing changes in the spike and other regions of the shell. (B) A hydrophobic pocket in the spike is marked by additional material in the DNA map, between the α-helical hairpins forming the spike. Mutations of the hydrophobic residues forming the pocket (P5, L60, L95, K96, I97) give rise to core proteins that do not become correctly enveloped and lead to defects in secretion of the virus.

What could be triggering the changes? During reverse transcription, single-stranded and therefore flexible RNA is being transcribed first into single-stranded DNA, with degradation of the RNA, and then into stiffer double-stranded DNA. There will be a corresponding change in charge density and nucleic acid flexibility that will have the effect of greatly increasing the force on the inner surface of the capsid. We believe that it is this increasing force that triggers the change in the core structure and enables envelopment to take place (Roseman et al., 2005). This is an example where structural observations have given an insight into function.

More recently, icosahedral viruses have once again been in the forefront of the steady advance toward atomic resolution. Studies include a 4.5-Å map of bacteriophage ε15, which showed the backbone structure of the capsid protein (Jiang et al., 2008); ~4 Å maps of the rotavirus double-layer particle (Zhang, X. et al., 2008) and of the triple-layer particle (Chen et al., 2009) which showed interactions important for assembly and uncoating; and a 3.3-Å map of a subviral particle of aquareovirus (Zhang, X. et al., 2010). The quality of these maps now matches or sometimes exceeds that from X-ray crystal structures of viruses, and these outstanding results will be reviewed in chapter 1 in Part B (Vol. 82) by Zhou.

IX. Single-Particle Analysis

I have concentrated on the analysis of particles with high symmetry, as this was where much of the significant progress had occurred. However, from fairly early on, a very important parallel development was occurring, stimulated partly by the desire to analyze particles, such as ribosomes, with no symmetry and partly by the need to characterize and quantify differences between images. In its basic form, the assumption is made that the images collected from the micrographs represent different projections of a single underlying structure, contaminated by noise. After being aligned, the set of images is subjected to multivariate statistical analysis (van Heel and Frank, 1981), which quantifies the significant variation between images and groups together with classes of similar images that can then be averaged. The averaged classes represent noise-reduced projections from which, if the view directions can be determined, a three-dimensional map can be calculated. The body of techniques so developed, initially using images of negatively stained particles, has come to be known as single-particle microscopy. Detailed reviews of the methods involved can be found in van Heel et al.

(2000) and Frank (2006a). The availability of various comprehensive software packages (e.g., SPIDER, Frank et al., 1981; IMAGIC, van Heel and Keegstra, 1981; EMAN, Ludtke et al., 1999), each developed over many years and based on these ideas, has been absolutely essential for progress in the analysis of cryo-images of particles with no or low symmetry. One of the highest resolution maps so far obtained (4.3 Å) is of a chaperonin with D8 symmetry (Zhang, J. et al., 2010), which demonstrates the outstanding results that can be achieved. However, many biologically important complexes have now been studied at medium resolution (10–20 Å), and this aspect of the subject is covered in Chapter 3.

The early assumption that the set of images represents views of a single state of the imaged structure has proved too restrictive for many macromolecular assemblies and machines. Sometimes, one part of the structure might protrude and be flexible, such that averaging images with the protrusion in different positions would lead to that feature being smeared out in the averaging process. Equally in a machine, such as the ribosome, large parts of the structure might be in different configurations at different points in the reaction pathway. In situations where the specimen contains a substrate or label, partial occupancy can again lead to heterogeneity in the image set. Analysis of the heterogeneity can frequently give insights into the function of the complex. Methods such as for determining where in the structure the variability lies (Zhang, W. et al., 2008) or for sorting and averaging the different states (Scheres et al., 2007) are therefore certain to play a crucial role in image analysis in the future. These matters are addressed in Chapter 4.

X. Tomography

In the extract from DeRosier and Klug (1968) quoted above in Section III, mention was made of the possibility of reconstruction from sections of biological material. For this kind of specimen, where each specimen is likely to contain a unique structure, as complete as possible a set of tilts must be collected for the reconstruction, a technique that has now come to be known as electron tomography. An early example of such a procedure with plastic sections of stained fish muscle used dual axis tilting for data collection but exploited the pseudocrystalline nature of the M-band to create an averaged image of the repeating unit (Luther and Crowther, 1984). As with negatively stained particulate specimens, the methods of preparation of plastic-sectioned stained material limited the amount of biologically relevant information that could be derived. However, the introduction of cryo-microscopy opened a new range of possibilities for tomography, including the imaging of whole cells, of frozen sections, or indeed of particulate material. The basic principles of tomography and many of the analytical methods developed are covered by the authors of the various chapters in Frank (2006b).

Cryo-tomography of particulate specimens can be useful for samples with low symmetry or for symmetrical samples that contain a nonsymmetrical part, as well as for samples that contain a range of structures. Herpes simplex virus (Grünewald et al., 2003) and HIV (Briggs et al., 2009) provide good examples of this kind of specimen. The first map of the clathrin cage involved tilt data collection and exploited the 622 symmetry of the hexagonal barrel form of cage to produce a map with a nominal resolution of about 50 Å (Vigers et al., 1986). The resolution was improved to 21 Å by single-particle methods (Smith et al., 1998). That a quasiatomic model of the clathrin cage has been made by fitting X-ray structures into an ~8 Å map (Fotin et al., 2004) emphasizes the huge progress that has been made with single-particle methods combined with component X-ray crystal structures. Single-particle tomography may itself be useful for creating an initial model for use in higher resolution single-particle structure determination. The current state of single-particle tomography will be described in chapter 2 in Part B (Vol. 82) by Schmid.

For tomography of whole cells, the size of eukaryotic cells means that the frozen sample is frequently too thick for good images to be collected, except possibly around the thinner edges of spread cells, for example, in *Dictyostelium* (Medalia et al., 2002). Thus, isolation of either subcellular organelles or cryo-sectioning is generally necessary to obtain samples of suitable thickness. However, prokaryotic cells are much smaller and here good progress has been made in visualizing the bacterial cytoskeleton in whole cells (e.g., Kürner et al., 2005). The general topic of cellular tomography is covered in chapter 3 in Part B (Vol. 82) by Hoenger. The main problems for cellular tomography are to establish the molecular identity of structures seen in tomograms and to home-in on interesting events at the molecular level in the vast space of the cell. To overcome these problems, it may be useful to identify and localize subcellular events by using fluorescent light microscopy and then to investigate the relevant region of the cell at molecular resolution with electron tomography. Such an approach will be the topic of chapter 4 in Part B (Vol. 82) by Hanein.

XI. Summary

I hope I have given an impression of the huge progress that has been made in the determination of biological structures using electron microscopy. Far more details will be found in the other chapters of this book. The whole field is at an immensely exciting stage, where many of the promises of the early ideas and developments have now been spectacularly fulfilled in a way that it was hard to imagine happening, even a few years ago. I was amused to find the report of an early workshop that I attended in 1973, entitled ‘‘The possibilities and prospects of obtaining high-resolution information (below 30 Å) on biological material using the electron microscope’’ (Beer et al., 1975). Despite the lack of ambition (below 30 Å), the whole tenor of the report is profoundly pessimistic, so it is probably a good thing that it has been largely ignored (cited only three times). Would the authors have believed that 35 years later we would have essentially atomic structures of two-dimensional crystals, helical assemblies, and icosahedral viruses determined by electron microscopy? And that the prospects for high-resolution maps of nonsymmetric particles and for detailed pictures of subcellular organization are just as promising?

Despite the advances, various problems still need to be addressed to enable cryo-EM to become a routine method of high-resolution structure determination. For example, in single-particle work, there is a need for structure validation methods to assess the reliability of the computed maps, such as might be achieved by an extension of the tilt pair method for investigating the accuracy of orientation determination (Rosenthal and Henderson, 2003). There is also a need for improvements in electronic image recording (Faruqi and McMullan, 2010), so as to achieve the optimal signal-to-noise ratio for transmission of high-resolution information that is needed for determining structures of lower molecular weight (Henderson, 1995). These and other matters will be addressed in chapter 5 in Part B (Vol. 82) by Rossmann on future prospects. But the field can be proud of its achievements in creating an approach that now rivals X-ray crystallography in its power for structure determination but that is applicable to a far wider range of biological objects.

References

Adrian, M., Dubochet, J., Lepault, J., McDowall, A. W. (1984). Cryo-electron microscopy of viruses. *Nature* **308**, 32–36.

Arndt, U. W., Crowther, R. A., Mallett, J. F. W. (1968). A computer linked cathode ray tube densitometer for X-ray crystallography. *J. Sci. Instrum.* **1**(2), 510–516.

Beer, M., Frank, J., Hanszen, K.-J., Kellenberger, E., Williams, R. C. (1975). The possibilities and prospects of obtaining high-resolution information (below 30 Å) on biological material using the electron microscope. *Q. Rev. Biophys.* **7**, 211–238.

Beroukhim, R., Unwin, N. (1997). Distortion correction of tubular crystals: improvements in the acetylcholine receptor structure. *Ultramicroscopy* **70**, 57–81.

Berriman, J. A., Serpell, L. C., Oberg, K. A., Fink, A. L., Goedert, M., Crowther, R. A. (2003). Tau filaments from human brain and from *in vitro* assembly of recombinant protein show cross-β structure. *Proc. Natl. Acad. Sci. USA* **100**, 9034–9038.

Böttcher, B., Wynne, S. A., Crowther, R. A. (1997). Determination of the fold of the core protein of hepatitis B virus by electron cryomicroscopy. *Nature* **386**, 88–91.

Brenner, S., Horne, R. W. (1959). A negative staining method for high resolution electron microscopy of viruses. *Biochim. Biophys. Acta* **34**, 103–110.

Briggs, J. A. G., Riches, J. D., Glass, B., Bartonova, V., Zanetti, G., Kräusslich, H.-G. (2009). Structure and assembly of immature HIV. *Proc. Natl. Acad. Sci. USA* **106**, 11090–11095.

Caspar, D. L. D., Klug, A. (1962). Physical principles in the construction of regular viruses. *Cold Spring Harb. Symp. Quant. Biol.* **27**, 1–23.

Chen, J. Z., Settembre, E. C., Aoki, S. T., Zhang, X., Bellamy, A. R., Dormitzer, P. R., et al. (2009). Molecular interactions in rotavirus assembly and uncoating seen by high-resolution cryo-EM. *Proc. Natl. Acad. Sci. USA* **106**, 10644–10648.

Cheng, R. H., Baker, T. S. (1996). A model-based approach for determining orientations of biological molecules imaged by cryoelectron microscopy. *J. Struct. Biol.* **116**, 120–130.

Cheng, R. H., Reddy, V. S., Olson, N. H., Fisher, A. J., Baker, T. S., Johnson, J. E. (1994). Functional implications of quasi-equivalence in a T=3 icosahedral animal virus established by cryo-electron microscopy and X-ray crystallography. *Structure* **2**, 271–282.

Conway, J. F., Cheng, N., Zlotnick, A., Winfield, P. T., Stahl, S. J., Steven, A. C. (1997). Visualization of a 4-helix bundle in the hepatitis B virus capsid by cryo-electron microscopy. *Nature* **386**, 91–94.

Crowther, R. A. (1971). Procedures for three-dimensional reconstruction of spherical viruses by Fourier synthesis from electron micrographs. *Phil. Trans. R. Soc. Lond. B* **261**, 221–230.

Crowther, R. A. (1991). Straight and paired helical filaments in Alzheimer disease have a common structural subunit. *Proc. Natl. Acad. Sci. USA* **88**, 2288–2292.

Crowther, R. A. (2004). Viruses and the development of quantitative biological electron microscopy. *Notes Rec. R. Soc. Lond.* **58**, 65–81.

Crowther, R. A. (2008). The Leeuwenhoek lecture 2006. Microscopy goes cold: frozen viruses reveal their structural secrets. *Phil. Trans. R. Soc. Lond. B* **363**, 2441–2451.

Crowther, R. A., Amos, L. A. (1971). Harmonic analysis of electron microscope images with rotational symmetry. *J. Mol. Biol.* **60**, 123–130.

Crowther, R. A., Amos, L. A., Finch, J. T., DeRosier, D. J., Klug, A. (1970). Three-dimensional reconstructions of spherical viruses by Fourier synthesis from electron micrographs. *Nature* **226**, 421–425.

Crowther, R. A., DeRosier, D. J., Klug, A. (1970). The reconstruction of a three-dimensional structure from projections and its application to electron microscopy. *Proc. R. Soc. Lond. A* **317**, 319–340.

Crowther, R. A., Kiselev, N. A., Böttcher, B., Berriman, J. A., Borisova, G. P., Ose, V., et al. (1994). Three-dimensional structure of hepatitis B virus core particles determined by electron cryomicroscopy. *Cell* **77**, 943–950.

DeRosier, D. J., Klug, A. (1968). Reconstruction of three dimensional structures from electron micrographs. *Nature* **217**, 130–134.

Dubochet, J., Adrian, M., Chang, J.-J., Homo, J.-C., Lepault, J., McDowall, A. W., et al. (1988). Cryo-electron microscopy of vitrified specimens. *Q. Rev. Biophys.* **21**, 129–228.

Egelman, E. H. (2000). A robust algorithm for the reconstruction of helical filaments using single particle methods. *Ultramicroscopy* **85**, 225–234.

Erickson, H. P., Klug, A. (1971). Measurement and compensation of defocusing and aberrations by Fourier processing of electron micrographs. *Phil. Trans. R. Soc. Lond. B* **261**, 105–118.

Faruqi, A. R., McMullan, G. (2010). Electronic detectors for electron microscopy. *Q. Rev. Biophys.* (In press).

Fotin, A., Cheng, Y., Sliz, P., Grigorieff, N., Harrison, S. C., Kirchhausen, T., et al. (2004). Molecular model for a complete clathrin lattice from electron cryomicroscopy. *Nature* **432**, 573–579.

Frank, J. (2006a). Three-Dimensional Electron Microscopy of Macromolecular Assemblies: Visualization of Biological Molecules in Their Native State. Oxford University Press, Inc., New York.

Frank, J. (2006b). Electron Tomography. Springer, New York.

Frank, J., Shimkin, B., Dowse, H. (1981). SPIDER—a modular software system for electron image processing. *Ultramicroscopy* **6**, 343–358.

Glaeser, R. M., Downing, K., DeRosier, D., Chiu, W., Frank, J. (2007). Electron Crystallography of Biological Macromolecules. Oxford University Press, Inc., New York.

Gonen, T., Cheng, Y., Sliz, P., Hiroaki, Y., Fujioshi, Y., Harrison, S. C., et al. (2005). Lipid-protein interactions in double-layered two-dimensional AQP0 crystals. *Nature* **438**, 633–638.

Grigorieff, N., Ceska, T. A., Downing, K. H., Baldwin, J. M., Henderson, R. (1996). Electron-crystallographic refinement of the structure of bacteriorhodopsin. *J. Mol. Biol.* **259**, 393–421.

Grünewald, K., Desai, P., Winkler, D. C., Heymann, J. B., Belnap, D. M., Baumeister, W., et al. (2003). Three-dimensional structure of herpes simplex virus from cryo-electron tomography. *Science* **302**, 1396–1398.

Hall, C. E. (1955). Electron densitometry of stained virus particles. *J. Biophys. Biochem. Cytol.* **1**, 1–15.

Henderson, R. (1995). The potential and limitations of neutrons, electrons and X-rays for atomic resolution microscopy of unstained biological molecules. *Q. Rev. Biophys.* **28**, 171–193.

Henderson, R., Baldwin, J. M., Ceska, T. A., Zemlin, F., Beckmann, E., Downing, K. H. (1990). Model for the structure of bacteriorhodopsin based on high resolution electron microscopy. *J. Mol. Biol.* **213**, 899–929.

Henderson, R., Unwin, P. N. T. (1975). Three-dimensional model of purple membrane obtained by electron microscopy. *Nature* **257**, 28–32.

Hoppe, W. (1970). Principles of structure analysis at high resolution using conventional electron microscopes and computers. *Berichte der Bunsen-Gesellschaft für Physicalische Chemie* **74**, 1090–1100.

Hoppe, W., Langer, R., Knesch, G., Poppe, C. (1968). Protein crystal structure analysis with electron rays. *Naturwissenschaften* **55**, 333–336.

Huxley, H. E. (1956). Some Observations on the Structure of Tobacco Mosaic Virus. First Eur. Conf. on Electron Microscopy, Stockholm, pp. 260–261.

Huxley, H. E. and Klug, A. (Eds.) (1971). New developments in electron microscopy. *Phil. Trans. Roy. Soc. Lond. B***261**.

Jiang, W., Baker, M. L., Jakana, J., Weigele, P. R., King, J., Chiu, W. (2008). Backbone structure of the infectious ε15 virus capsid revealed by electron cryomicrocopy. *Nature* **451**, 1130–1134.

Klug, A., Berger, J. E. (1964). An optical method for the analysis of periodicities in electron micrographs, and some observations on the mechanism of negative staining. *J. Mol. Biol.* **10**, 565–569.

Klug, A., Crick, F. H. C., Wykoff, H. W. (1958). Diffraction by helical structures. *Acta Cryst.* **11**, 199–213.

Klug, A., DeRosier, D. J. (1966). Optical filtering of electron micrographs: reconstruction of one-sided images. *Nature* **212**, 29–32.

Klug, A., Finch, J. T. (1965). Structure of viruses of the papilloma–polyoma type I. Human wart virus. *J. Mol. Biol.* **11**, 403–423.

Klug, A., Finch, J. T. (1968). Structure of viruses of the papilloma–polyoma type. IV. Analysis of tilting experiments in the electron microscope. *J. Mol. Biol.* **31**, 1–12.

Kürner, J., Frangakis, A. S., Baumeister, W. (2005). Cryo-electron tomography reveals the cytoskeletal structure of *Spiroplasma melliferum. Science* **307**, 436–438.

Low, H. H., Sachse, C., Amos, L. A., Löwe, J. (2009). Structure of a dynamin-like protein lipid tube provides a mechanism for assembly and membrane curving. *Cell* **139**, 1342–1352.

Ludtke, S. J., Baldwin, P. R., Chiu, W. (1999). EMAN: semiautomated software for high-resolution single-particle reconstructions. *J. Struct. Biol.* **128**, 82–97.

Luther, P. K., Crowther, R. A. (1984). Three-dimensional reconstruction from tilted sections of fish muscle M-band. *Nature* **307**, 566–568.

Markham, R., Frey, S., Hills, G. J. (1963). Methods for the enhancement of image detail and accentuation of structure in electron microscopy. *Virology* **20**, 88–102.

Markham, R., Hitchborn, J. H., Hills, G. J., Frey, S. (1964). The anatomy of the tobacco mosaic virus. *Virology* **22**, 342–359.

Medalia, O., Weber, I., Frangakis, A. S., Nicastro, D., Gerisch, G., Baumeister, W. (2002). Macromolecular architecture in eukaryotic cells visualized by cryoelectron tomography. *Science* **298**, 1209–1213.

Miyazawa, A., Fujiyoshi, Y., Unwin, N. (2003). Structure and gating mechanism of the nicotinic acetylcholine receptor at 4 Å resolution. *Nature* **423**, 949–955.

Roseman, A. M., Berriman, J. A., Wynne, S. A., Butler, P. J. G., Crowther, R. A. (2005). A structural model for maturation of the hepatitis B virus core. *Proc. Natl. Acad. Sci. USA* **102**, 15821–15826.

Rosenthal, P. B., Henderson, R. (2003). Optimal determination of particle orientation, absolute hand, and contrast loss in single-particle electron microscopy. *J. Mol. Biol.* **333**, 721–745.

Ruska, E. (1986). The development of the electron microscope and of electron microscopy. Available at www.nobel.se/laureates/1986/ruska-lecture.pdf [Nobel lecture].

Sachse, C., Chen, J. Z., Coureux, P.-D., Stroupe, M. E., Fändrich, M., Grigorieff, N. (2007). High-resolution electron microscopy of helical specimens: a fresh look at tobacco mosaic virus. *J. Mol. Biol.* **371**, 812–835.

Sachse, C., Fändrich, M., Grigorieff, N. (2008). Paired β-sheet structure of an Aβ(1-40) amyloid fibril revealed by electron microscopy. *Proc. Natl. Acad. Sci. USA* **105**, 7462–7466.

Sawaya, M. R., Sambashivan, S., Nelson, R., Ivanova, M. I., Sievers, S. A., Apostol, M. I., et al. (2007). Atomic structures of amyloid cross-beta spines reveal varied steric zippers. *Nature* **447**, 453–457.

Scheres, S. H. W., Gao, H., Valle, M., Herman, G. T., Eggermont, P. P. B., Frank, J., et al. (2007). Disentangling conformational states of macromolecules in 3D-EM through likelihood optimization. *Nat. Methods* **4**, 27–29.

Serpell, L. C., Berriman, J. A., Jakes, R., Goedert, M., Crowther, R. A. (2000). Fibre diffraction of synthetic α-synuclein filaments shows amyloid-like cross-β structure. *Proc. Natl. Acad. Sci. USA* **97**, 4897–4902.

Smith, C. J., Grigorieff, N., Pearse, B. M. F. (1998). Clathrin coats at 21 Å resolution: a cellular assembly designed for multiple membrane receptors. *EMBO J.* **17**, 4943–4953.

Taylor, K. A., Glaeser, R. M. (1974). Electron diffraction of frozen, hydrated protein crystals. *Science* **186**, 1036–1037.

Unwin, N. (2005). Refined structure of the nicotinic acetylcholine receptor at 4 Å resolution. *J. Mol. Biol.* **346**, 967–989.

Unwin, P. N. T., Henderson, R. (1975). Molecular structure determination by electron microscopy of unstained crystalline specimens. *J. Mol. Biol.* **94**, 425–440.

van Heel, M., Frank, J. (1981). Use of multivariate statistics in analyzing the images of biological macromolecules. *Ultramicroscopy* **6**, 187–194.

van Heel, M., Gowen, B., Matadeen, R., Orlova, E. V., Finn, R., Pape, T., et al. (2000). Single-particle electron cryo-microscopy: towards atomic resolution. *Q. Rev. Biophys.* **33**, 307–369.

van Heel, M., Keegstra, W. (1981). IMAGIC: a fast, flexible and friendly image analysis software system. *Ultramicroscopy* **7**, 113–130.

Vigers, G. P. A., Crowther, R. A., Pearse, B. M. F. (1986). Three dimensional structure of clathrin cages in ice. *EMBO J.* **5**, 529–534.

Wynne, S. A., Crowther, R. A., Leslie, A. G. W. (1999). Crystal structure of the human hepatitis B virus capsid. *Mol. Cell* **3**, 771–780.

Yonekura, K., Maki-Yonekura, S., Namba, K. (2003). Complete atomic model of the bacterial flagellar filament by electron cryomicroscopy. *Nature* **424**, 643–650.

Zhang, J., Baker, M. L., Schröder, G. F., Douglas, N. R., Reissmann, S., Jakana, J., et al. (2010). Mechanism of folding chamber closure in a group II chaperonin. *Nature* **463**, 379–383.

Zhang, X., Jin, L., Fang, Q., Hui, W. H., Zhou, Z. H. (2010). 3.3 Å cryo-EM structure of a nonenveloped virus reveals a priming mechanism for cell entry. *Cell* **141**, 472–482.

Zhang, W., Kimmel, M., Spahn, C. M. T., Penczek, P. A. (2008). Heterogeneity of large macromolecular complexes revealed by 3D cryo-EM variance analysis. *Structure* **16**, 1770–1776.

Zhang, X., Settembre, E., Xu, C., Dormitzer, P. R., Bellamy, R., Harrison, S. C., et al. (2008). Near-atomic resolution using electron cryomicroscopy and single-particle reconstruction. *Proc. Natl. Acad. Sci. USA* **105**, 1867–1872.

PRESENT AND FUTURE OF MEMBRANE PROTEIN STRUCTURE DETERMINATION BY ELECTRON CRYSTALLOGRAPHY

By IBAN UBARRETXENA-BELANDIA* AND DAVID L. STOKES†,‡

*Department of Structural and Chemical Biology, Mt. Sinai School of Medicine, New York, New York, USA
†Skirball Institute and Department of Cell Biology, New York University School of Medicine, New York, New York, USA
‡New York Structural Biology Center, Division of Cryo-Electron Microscopy, New York, New York, USA

ADVANCES IN PROTEIN CHEMISTRY AND STRUCTURAL BIOLOGY, Vol. 81
DOI: 10.1016/S1876-1623(10)81002-8

Abstract

Membrane proteins are critical to cell physiology, playing roles in signaling, trafficking, transport, adhesion, and recognition. Despite their relative abundance in the proteome and their prevalence as targets of therapeutic drugs, structural information about membrane proteins is in short supply. This chapter describes the use of electron crystallography as a tool for determining membrane protein structures. Electron crystallography offers distinct advantages relative to the alternatives of X-ray crystallography and NMR spectroscopy. Namely, membrane proteins are placed in their native membranous environment, which is likely to favor a native conformation and allow changes in conformation in response to physiological ligands. Nevertheless, there are significant logistical challenges in finding appropriate conditions for inducing membrane proteins to form two-dimensional arrays within the membrane and in using electron cryo-microscopy to collect the data required for structure determination. A number of developments are described for high-throughput screening of crystallization trials and for automated imaging of crystals with the electron microscope. These tools are critical for exploring the necessary range of factors governing the crystallization process. There have also been recent software developments to facilitate the process of structure determination. However, further innovations in the algorithms used for processing images and electron diffraction are necessary to improve throughput and to make electron crystallography truly viable as a method for determining atomic structures of membrane proteins.

I. Introduction

Biological membranes surround all cells and mediate all their interactions with the outside world. Membrane proteins relay information or chemical substrates across the membrane and are key players in the biochemical events that take place either at the surface of cells or within membrane-bound organelles. Depending on the biological context, membrane proteins act as receptors, enzymes, channels, transporters, structural proteins, and cell–cell adhesion molecules and, as such, contribute to an astounding variety of essential cellular functions, including transmembrane signaling, homeostasis, and energy conversion.

When considered on a genome-wide scale, membrane proteins comprise ~40% of all genes in eukaryotic, eubacterial, and archaeal organisms (Wallin and von Heijne, 1998). Given their omnipresence and functional diversity, it is not surprising that membrane proteins play a pivotal role in numerous human pathologies. Important diseases resulting from defective membrane proteins include cystic fibrosis, several forms of cancer, Alzheimer's disease, and various cardiomyopathies. In fact, ~60% of the therapeutic drugs currently used in the United States target membrane proteins (Drews, 2000).

Despite this tremendous relevance to basic cell biology and to therapeutic medicine, our understanding of membrane proteins from a structural perspective is limited, especially when it comes to visualizing membrane proteins in their natural lipid bilayer environment. To a large extent, this limitation is due to the prevalent tools for structure determination: X-ray crystallography and NMR spectroscopy. These tools have become increasingly successful with detergent-solubilized species, but are at a distinct disadvantage when membrane proteins are embedded in a lipid bilayer. In contrast, electron crystallography is particularly well suited to the study of membrane proteins within their native, membranous environment. Although electron crystallography provided the first three-dimensional (3D) structure of a membrane protein—bacteriorhodopsin in 1975—and has subsequently produced a handful of atomic structures, it has largely foundered in the fringes of structural biology due to practical difficulties and lack of a high-throughput approach. Here, we review its current state-of-the-art and discuss the future developments that are necessary to allow electron crystallography to fulfill its promise.

II. The Membrane Environment

Lipid molecules are arranged as a continuous bimolecular layer of approximately 50–60 Å in thickness (Engelman, 1971; Mitra et al., 2004). The lipid bilayer is a structurally and chemically heterogeneous environment that complicates the surface chemistry of membrane proteins relative to their soluble cousins. Three distinct regions can be delineated in a cross section of the bilayer: (1) the hydrophobic core populated by the lipid acyl chains, (2) the hydrophilic layers flanking the core that are formed by lipid head groups, and (3) the aqueous regions at the outer margins (White and Wimley, 1999). The hydrophobic core is ~30 Å thick and largely impermeable to polar molecules and ions. This is a region with a low dielectric that favors long-range polar interactions and where the hydrophobic effect is absent. The length of the acyl chains and their degree of saturation influence the overall thickness and fluidity of the bilayer. The lipid head groups occupy 10–15 Å on either side of the hydrophobic core and serve to bind most of the water in these regions. As a result, the hydration of protein components and the magnitude of the hydrophobic effect are decreased compared with bulk aqueous solution. Lipids contain a variety of different head groups, which can include charge, dipoles, and carbohydrate groups. Finally, the surrounding aqueous environment in the vicinity of most membranes also displays distinct properties relative to bulk water. Due mainly to surface charge from the lipid head groups, this region typically displays a gradient in solutes, pH, and ions. As an additional complexity, the lipids composing biological membranes are heterogeneous (Brugger et al., 1997), differing between the two leaflets of the bilayer and between membrane compartments within a given cell (Pike et al., 2002). In this way, the cell can adjust the thickness, surface charge, and fluidity of membranes to meet the requirements of individual membrane proteins in different cellular compartments (Yeagle, 1989; Dowhan, 1997; Lee, 2004; Andersen and Koeppe, 2007; Nyholm et al., 2007).

In order to conform to their membrane environment, integral membrane proteins are amphiphilic in nature, with their transmembranous regions immersed in the hydrophobic core of a lipid bilayer and their extramembranous domains surrounded by water. The water-exposed domains adopt the diverse array of protein folds observed in soluble proteins, though their vicinity to the membrane surface likely influences their design. The structure of membrane domains is dictated by the physical and chemical constraints of the lipid bilayer (White and Wimley, 1999; Popot and Engelman, 2000; Schulz, 2000; Ubarretxena-Belandia and Engelman, 2001), and membrane protein structures so far reveal membrane domains composed either of α-helical bundles or β-barrels. β-barrel architectures are largely constrained to the outer membrane of bacteria, and in this chapter, we will focus on α-helical membrane proteins which have a greater influence over the functioning of eukaryotic cells and human tissues, in particular.

III. Why Are There so Few Membrane Protein Structures?

This amphiphilic nature of membrane proteins represents a fundamental constraint on our ability to produce and to study these proteins. Specifically, there are serious hurdles associated with (1) overexpression of membrane proteins with native tertiary and quaternary structure; (2) preservation of biological activity when membrane proteins are extracted from their native membrane environment with detergent; (3) the large size of membrane protein/detergent complexes, which limits the application of solution NMR; and (4) the difficulty in obtaining well-diffracting 3D crystals for X-ray crystallography. These obstacles become more pronounced for large membrane protein complexes or for the less stabile membrane proteins that tend to come from eukaryotic sources. The limited capacity of the cell to accommodate overexpression is due either to the physiological consequences on membrane function or to the limited capacity to produce extra membrane surface area. Limited stability reflects the inability of detergent to duplicate the physical/chemical environment of the lipid bilayer; although the basic tripartite structure is present in a lipid micelle, the heterogeneity, lateral pressure, and charge distribution of a biological membrane are impossible to replicate. Finally, 3D crystallization relies primarily on intermolecular contacts between extramembranous regions of the protein and conditions for promoting these interactions while simultaneously stabilizing the intramembranous region of the protein are difficult to find.

Our inability to overcome these hurdles is clearly reflected in the Protein Data Bank (PDB): out of ~63,000 protein structures deposited as of September 2010, only ~691 structures come from 256 different membrane proteins (http://blanco.biomol.uci.edu/Membrane_Proteins_xtal.html). The β-barrel fold is greatly overrepresented (25%) in the PDB, reflecting the enhanced stability of this fold and the ease of producing proteins from bacterial hosts. Nevertheless, there has been tremendous progress over the last several years and the number of membrane protein structures has started to increase exponentially (White, 2009). Recent success stories include a number of groundbreaking structures that have greatly illuminated their respective fields, creating an increased appetite for improving existing technologies and for finding new and better ways to study the structures of this vast class of biological macromolecules. By increasing our understanding of membrane protein structure and, specifically, by appreciating the influence of their native lipid environment on their function, we will be better equipped to understand the role of these proteins in human health and disease.

IV. Application of Electron Crystallography to Membrane Proteins

Electron crystallography is the only method capable of imaging membrane proteins in their lipid environment. This method was pioneered in the 1970s by Henderson and Unwin in their studies of bacteriorhodopsin (Henderson and Unwin, 1975) and relies on an ordered array of molecules within a bilayer in the form of two-dimensional (2D) sheets or tubes (Fig. 1). In the case of bacteriorhodopsin, ordering occurs *in vivo* on a specialized photosynthetic region of the plasma membrane, providing an ideal specimen to drive development of the necessary technologies. After significant advances in electron microscope design, imaging strategies, and image reconstruction algorithms, the atomic structure of bacteriorhodopsin was published in 1990 (Henderson et al., 1990), just 6 years after the landmark X-ray crystallographic structure of photosynthetic reaction center (Deisenhofer et al., 1984). Based primarily on these early developments by Henderson and colleagues, electron crystallography has continued to be a powerful tool for studying 3D structure of membrane proteins (Table I) at medium and high resolution (Stahlberg et al., 2001; Unger, 2001; Subramaniam et al., 2002). In addition to bacteriorhodopsin, this methodology has yielded atomic structures of plant light-harvesting complex (Kühlbrandt et al., 1994), human red cell aquaporin-1 (Murata et al., 2000), eye lens aquaporin-0 (Gonen et al., 2005), rat aquaporin-4 (Hiroaki et al., 2006), glutathione transferase (Holm et al., 2006), prostaglandin E synthase (Jegerschold et al., 2008) and acetylcholine receptor (Unwin, 2005). In addition, 3D structures of ~25 other unique membrane proteins have been determined to medium resolution (5–8 Å), and continuing efforts are expected to produce atomic models in the near future (e.g., Hirai et al., 2002; Ubarretxena-Belandia et al., 2003; Kukulski et al., 2005). The recent structure of aquaporin-0 (AQP0; Fig. 2) is noteworthy and deserves further mention, not only for its remarkably high resolution (1.9 Å) but also for its unique ability to reveal essentially all the lipid molecules that make up the membrane bilayer (Gonen et al., 2005). Thus, despite recent advances in the application of solid state NMR (Hong, 2006)

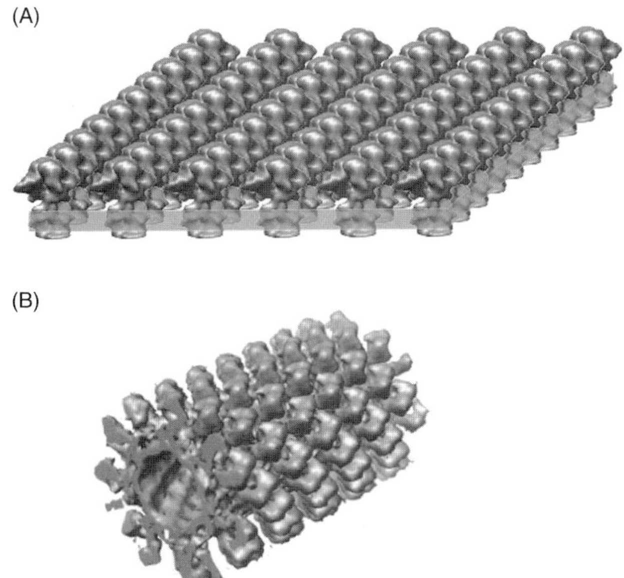

FIG. 1. Types of crystals used for membrane protein structure determination by electron crystallography. (A) A planar bilayer (highlighted as a gray slab) with a coherent 2D array of proteins. These crystals must be tilted to collect data for a 3D analysis of their structure. Another related type of crystal (not shown) arises from flattened lipid vesicles that contain two overlapping 2D lattices. (B) A helical array of proteins in a cylindrical lipid vesicle. Because many different views are provided for the molecules, these helical crystals do not need to be tilted. In both cases, crystals are very thin (50 Å for single-layered bacteriorhodopsin crystals and 600 Å for tubular crystals of the nicotinic acetylcholine receptor) and, as a result, cryo-EM combined with image processing is the natural choice for solving their 3D structure.

and molecular dynamics simulations (Lindahl and Sansom, 2008), electron crystallography represents the best approach to understanding membrane protein structure in the context of a lipid bilayer.

In special cases, crystallization within the lipid bilayer can be achieved directly within the native cellular membrane: for example, bacteriorhodopsin from *Halobacterium halobium*, Ca^{2+}-ATPase from mammalian sarcoplasmic reticulum (Zhang et al., 1998), and acetylcholine receptor from the electric organ of *Torpedo marmorata* (Unwin, 2005). More generally, these crystals are grown by reconstitution of purified, detergent-solubilized membrane proteins into lipid bilayers under defined conditions (for reviews, see

TABLE I
3D Structures of Membrane Proteins Determined by Electron Crystallography

Membrane protein[a]	Resolution (Å)	Year	Reference
Eye lens aquaporin 0	1.9	2005	Gonen et al. (2005)
Aquaporin-4	2.8	2009	Tani et al. (2009)
Bacteriorhodopsin	3.0	1997	Kimura et al. (1997)
Glutathione transferase	3.2	2006	Holm et al. (2006)
Plant LHC-II	3.4	1994	Kühlbrandt et al. (1994)
Bacteriorhodopsin	3.5	1990	Henderson et al. (1990)
Prostaglandin E synthase	3.5	2008	Jegerschold et al. (2008)
Aquaporin-1	3.8	2000	Murata et al. (2000)
Acetylcholine receptor	4.0	2005	Unwin (2005)
Human aquaporin 2	4.5	2005	Schenk et al. (2005)
Plant aquaporin SoPIP2	5.0	2005	Kukulski et al. (2005)
Halorhodopsin	5.0	2000	Kunji et al. (2000)
Bovine rhodopsin	5.5	2003	Krebs (2003)
Porin PhoE	6.0	1991	Jap et al. (1991)
Bacteriorhodopsin	6.0	1975	Henderson and Unwin (1975)
Glutathione transferase	6.0	2002	Holm et al. (2002)
Bacteriorhodopsin	6.5	1983	Leifer and Henderson (1983)
Oxalate transporter OxlT	6.5	2002	Hirai et al. (2002)
Frog rhodopsin frog	6.5	1997	Unger et al. (1997)
Ca^{2+}-ATPase	6.5	2002	Xu et al. (2002)
Glycerol channel GlpF	6.9	2000	Stahlberg et al. (2000)
Gap junction channel	7.0	2007	Oshima et al. (2007)
NhaA Na/ H^+ antiporter	7.0	2000	Williams (2000)
EmrE multidrug transporter	7.0	2003	Ubarretxena-Belandia et al. (2003)
hCTR1 Cu transporter	7.0	2009	De Feo et al. (2009)
Gap junction channel	7.5	1999	Unger et al. (1999)
Sec YEG complex	8.0	2005	Bostina et al. (2005)
Plant photosystem II RC	8.0	1998	Rhee et al. (1998)
Neurospora H^+-ATPase	8.0	1998	Auer et al. (1998)
Acetylcholine receptor	9.0	1993	Unwin (1993)

[a]Atomic-resolution structures in bold.

Jap et al., 1992; Kühlbrandt, 1992; Mosser, 2001). Reconstitution involves the controlled removal of detergent—by dialysis (Kühlbrandt, 1992), by controlled dilution (Remigy et al., 2003), by adsorption onto a hydrophobic resin (Rigaud et al., 1997), or by complexation with

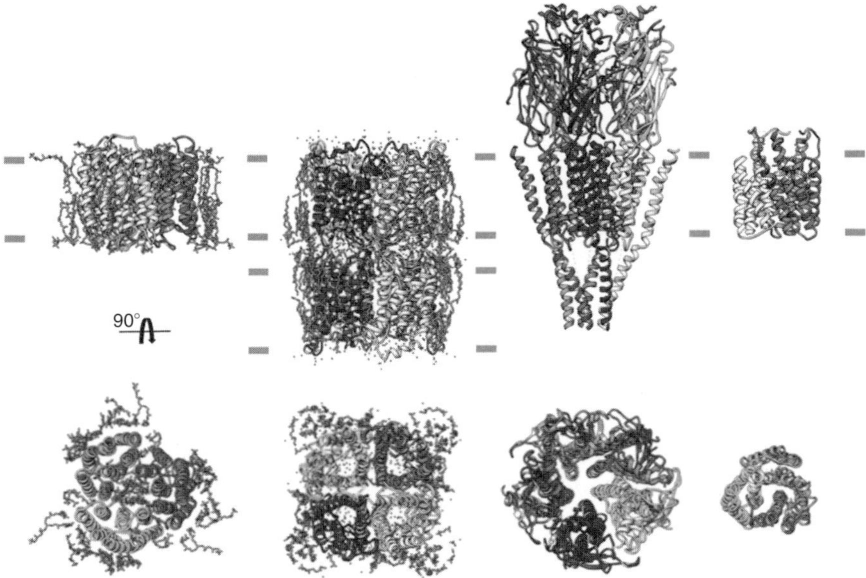

FIG. 2. Selection of high-resolution membrane protein structures solved by electron crystallography. From left to right: a trimer of bacteriorhodopsin, a double-layer of the tetrameric AQP0, the heteropentameric acetylcholine receptor, and a trimer of gluthathione transferase 1. The approximate boundaries of the bilayer are indicated by short blue lines, and individual lipid molecules present in the structure are shown in green. (See color plate 1).

cyclodextrins (Signorell et al., 2007)—in the presence of defined lipid species at an optimal lipid-to-protein ratio (LPR). By constraining a high density of a single protein species within a planar lipid bilayer, formation of a regular array within this bilayer becomes relatively favorable (Fig. 1). Given the physical constraint of molecules within the 2D plane of the bilayer, precipitants are not generally required for crystallization. Rather the most important factors appear to be the structural integrity and homogeneity of the protein, the choice of lipid species, the density of the protein within the bilayer, and the surface charge (as controlled by pH and lipid head group composition).

There are three predominant morphologies adopted by the resulting crystals: (1) flattened lipid vesicles with two, overlapping 2D lattices; (2) tubular vesicles which retain a cylindrical shape and contain a helically organized array of membrane proteins; and (3) a single, flat bilayer with a single, coherent 2D array of proteins (Fig. 1). Because these crystals are very thin (50 Å for single-layered bacteriorhodopsin crystals and 600 Å for tubular crystals of the nicotinic acetylcholine receptor), electron cryomicroscopy (cryo-EM) combined with image processing is the natural choice for solving their 3D structure.

V. Atomic Structures by Electron Crystallography

A. *Aquaporin*

Water is the most prevalent molecule in biological tissues and life could not exist without its free circulation into and out of cells. Water can diffuse spontaneously across cell membranes but only at very low rates, and for this reason, organisms from all three domains of life, from the simplest unicellular organisms to mammals, express membrane proteins called aquaporins that form specialized pores for water (Preston et al., 1992; Agre et al., 1993). There are many different types of aquaporins, at least 13 in mammals and at least five subfamilies in plants. Many of the mammalian aquaporins are expressed in the kidney, where water resorption is of critical physiological importance. Nevertheless, the original discovery of aquaporin was in the erythrocyte (AQP1), which set in motion a fierce competition to obtain the structure. Two different groups used electron crystallography to determine the first structures at 6–7 Å resolution (Cheng et al., 1997; Walz et al., 1997), which after several more years were improved to 3.8 Å (Murata et al., 2000), at which level the polypeptide chain could be traced. These structures revealed the protein fold for the first time and illustrated the path of water conduction across the membrane. Subsequent structures of aquaporins by electron crystallography include AQP2 (4.5 Å; Schenk et al., 2005) from apical membranes in the kidney collecting duct, AQP4 (2.8 Å; Tani et al., 2009) from basolateral membranes, plant aquaporin (SoPIP2 at 5.0 Å; Kukulski et al., 2005), and AQP0 from the eye lens at 2.5–1.9 Å resolution (Gonen et al., 2005; Hite et al., 2010).

The structures of AQP0 are notable for several reasons. First, this work represented the first time that methods of molecular replacement were applied in electron crystallography and the 1.9 Å resolution represented a new record for a mammalian membrane protein. Second, AQP0 formed double-layered crystals, which reflect the cell–cell junctions mediated by these channels in the lens (Fig. 2). Details of the structure showed how AQP0 closes its water channel upon formation of these junctions, a behavior that is important to maintaining proper hydration levels in the eye. Finally, these structures revealed a continuous lipid bilayer surrounding AQP0, elucidating specific interactions between the protein and its lipids. Furthermore, crystals formed from a different lipid species comprised a different oligomeric state of AQP0 (Hite et al., 2010), illustrating the strong effect that the lipid environment can have on membrane protein structure.

B. *Acetylcholine Receptor*

Synaptic transmission at the neuromuscular junction is mediated by acetylcholine. The postsynaptic membrane is therefore densely packed with the large heteropentameric acetylcholine receptor that recognizes acetylcholine and opens an ion channel. This opening depolarizes the membrane thus leading to an action potential and initiation of muscle contraction. A series of structures have been determined from tubular crystals of this nicotinic acetylcholine receptor from the electric organ of *T. marmorata* (Fig. 2). Unlike the aquaporins, these crystals form spontaneously within the native biological membrane and curvature is induced by the crystal contacts, which leads to closed cylindrical shape with helical symmetry. Due to their limited size, these helical crystals harbor relatively few molecules and thus produce a lower signal-to-noise ratio. The march to atomic resolution started modestly in 1981 with a structure at 30 Å resolution (Kistler and Stroud, 1981; Brisson and Unwin, 1984). Through a dogged improvement of imaging conditions and image processing algorithms (Beroukhim and Unwin, 1997), the resolution of the 3D structure gradually improved to 17 Å (Toyoshima and Unwin, 1990), 9 Å (Unwin, 1993), 4.6 Å (Miyazawa et al., 1999), and ultimately to 4 Å (Miyazawa et al., 2003), where an atomic model was built. High-resolution details were modeled with reference to X-ray crystallographic structures of a soluble acetylcholine binding protein, which forms a pentameric structure related to the cytoplasmic domains (Brejc et al., 2001), and to a bacterial homologue for the entire acetylcholine receptor (Hilf and Dutzler, 2008). In addition, the crystals of the nicotinic acetylcholine receptor have been used for studying the mechanism of gating, by spraying acetylcholine onto the sample prior to rapid (millisecond time scale) freezing (Berriman and Unwin, 1994; Unwin, 1995). The fact that these crystals provide the native membrane environment and that the constituent acetylcholine receptor molecules represent the heteropentameric assembly present at mammalian neuromuscular junction means that these studies provide invaluable insight that could not be obtained by X-ray crystallography.

C. *Glutathione Transferase*

Aquaporin 0, the nicotinic acetylcholine receptor and bacteriorhodopsin are all abundant in their native cellular membranes, the latter forming crystals directly within the membrane of *H. halobium.* Thus, it is perhaps not surprising that these proteins can form large 2D arrays with long-range crystalline order which diffract to atomic resolution. But what about the vast majority of membrane proteins, which are present only at low to moderate concentration in their biological membranes? Can they also form membrane crystals that diffract to high resolution? Microsomal glutathione transferase 1 is present at low levels in eukaryotic membranes, yet forms large membrane crystals that diffract to high resolution. This protein belongs to the superfamily of membrane-associated proteins in eicosanoid and glutathione metabolism. These proteins are key for synthesizing mediators of fever, pain, and inflammation as well as for protection against reactive molecules and oxidative stress. The structure of the rat microsomal glutathione transferase 1 has been recently solved at 3.2 Å resolution in complex with glutathione by electron crystallography (Holm et al., 2006) and the related prostaglandin E synthase has been solved at 3.5 Å resolution (Jegerschold et al., 2008). These proteins form a homotrimer (Fig. 2) and the former structure revealed a binding site for glutathione that differed from the canonical soluble glutathione transferases.

VI. Advantages of Membrane Crystals

A. *Crystallization Within the Membrane Requires Only Moderate Protein Concentrations*

Like soluble proteins, crystallization of detergent-solubilized membrane proteins for X-ray crystallography involves a phase transition that is facilitated by high protein concentrations (5–20 mg/ml). NMR also requires high concentrations of noncrystalline material to ensure suitable signal. In contrast, concentrations of 0.5–1 mg/ml are sufficient for reconstitution and crystallization of membrane proteins within the membrane bilayer. This is an important consideration for eukaryotic membrane proteins, which generally have low expression levels and a higher tendency to aggregate at higher concentrations. This tendency may be due to their exposure to very high detergent concentrations after concentration, or due to increased interactions between hydrophobic surfaces. A related benefit is that membrane crystals do not generally rely on a precipitating agent, such as high salt or polyethylene glycol. Such agents create nonphysiological conditions in the aqueous phase that can lead to precipitation or unnatural conformations.

B. The Membrane Environment Favors Native Protein Conformation

A protein structure is far more informative if it represents a physiological conformation. In the case of membrane proteins, the inhomogeneous dielectric, charge distribution, and lateral pressure of the bilayer can represent significant factors in determining the conformation and even the overall fold. In this regard, even lower resolution structures obtained from membrane crystals can yield valuable insights. For example, the X-ray structures of the EmrE multidrug transporter (Ma and Chang, 2004) caused considerable controversy which was resolved by electron crystallography (Ubarretxena-Belandia et al., 2003). EmrE is a multidrug transporter that catalyzes the electrogenic efflux of various cationic aromatic hydrocarbons in exchange for two protons. The 3D structure of EmrE was first determined at 7 Å by electron crystallography (Fig. 3) and showed a bundle of eight transmembrane α-helices with one substrate molecule (tetraphenylphosphonium) bound near the center (Ubarretxena-Belandia et al., 2003). The most remarkable finding was that EmrE formed an asymmetric homodimer with the two monomers related by a 180° rotation about an axis parallel to the membrane. This finding suggested that EmrE monomers are inserted with opposite topologies into the membrane. This antiparallel dimer represented a novel packing arrangement never before observed in a membrane protein. Subsequent X-ray structures of detergent-solubilized EmrE (Ma and Chang, 2004; Pornillos et al., 2005) revealed a completely different packing of transmembrane helices, the physiological relevance of which was challenged in light of the electron crystallographic structure (Fleishman et al., 2006). As a result, the X-ray structures were ultimately revised to become more consistent

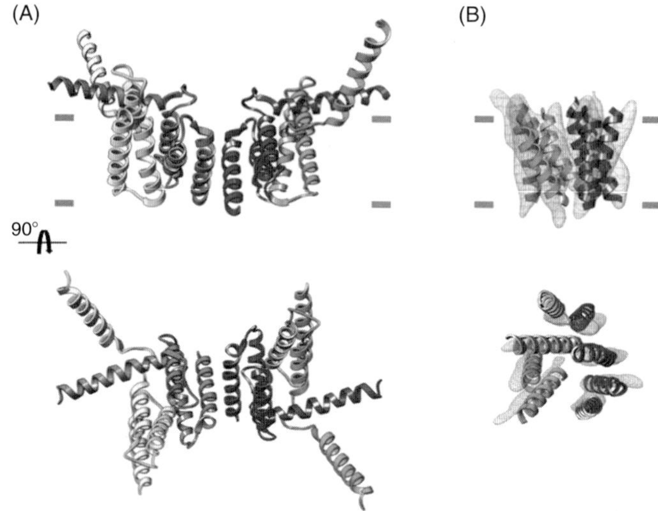

FIG. 3. Membrane crystals preserve the native structure of membrane proteins. (A) The original X-ray structure showing a dimer of the multidrug-resistance antiporter from *E. coli* EmrE solved from 3D crystals formed with detergent-solubilized protein (pdb 1S7B). (B) The electron crystallographic map of monomeric EmrE (emdb emd-1087) solved at a resolution of 7 Å fitted with the model derived by Fleishman et al. (2006) (pdb 2I68). In the X-ray structure, two of the helices are protruding radically from the presumed bilayer plane, which is indicated by the short blue lines. (See color plate 2).

with electron crystallographic data, showing the details of the antiparallel packing interaction at atomic resolution (Chen et al., 2007).

C. *Conformational Changes Are More Readily Accommodated in Membrane Crystals*

The physical constraints within membrane crystals are fewer than those for a 3D crystal, because the intermolecular interactions occur mainly in the 2D plane of the lipid bilayer rather than propagating isotropically in all three dimensions. Indeed, electron crystallographic studies of nicotinic acetylcholine receptor, rhodopsin, and the Na^+/H^+ antiporter from *Escherichia coli* (NhaA) have all included conformational changes induced by applying physiologically relevant stimuli to the membrane crystals. In the case of rhodopsin (reviewed in Schertler, 2005), light causes isomerization of 11-*cis* retinal, thus initiating the photoactivation process, which involves an equilibrium between meta-I and meta-II conformations. Rhodopsin is also photoactive in 3D crystals, but X-ray diffraction from these crystals deteriorates dramatically after illumination, presumably due to a disordering caused by the corresponding conformational change. In contrast, membrane crystals accommodate these structural changes and have allowed the photocycle of this light-activated proton pump to be characterized by electron crystallography (Ruprecht et al., 2004). A similar approach was undertaken with the bacterial light-activated proton pump, bacteriorhodopsin, where electron crystallography has been instrumental in studying conformational changes not tolerated by 3D crystals (Hirai and Subramaniam, 2009).

In the case of NhaA, pH changes were used to study the transport cycle. The initial electron crystallographic map of NhaA at 7 Å (Williams, 2000) and the ensuing atomic structure by X-ray crystallography (Hunte et al., 2005) were both obtained at pH 4, that is, with transport sites saturated with protons. To obtain mechanistic insight into other conformations of NhaA, NhaA membrane crystals were soaked in different buffers and electron crystallography revealed two pH activated states of the transporter (Appel et al., 2009), an approach that has not yet been possible with the 3D crystals.

D. *Membrane Crystals Offer an Optimal Binding Surface for Aqueous Ligands*

The physiological topology of membrane proteins within membrane crystals exposes extramembranous surfaces to the aqueous medium and the high density of proteins within the crystal ameliorates ''low occupancy'' that can be obtained with ligands with limited affinity. This property has been exploited to study the conformational changes of the multidrug transporter EmrE in response to ligands of varying sizes (Tate et al., 2003; Ubarretxena-Belandia and Tate, 2004; Korkhov and Tate, 2008).

VII. Current Hurdles in Electron Crystallography

Although it is clear from these examples that electron crystallography is well suited for structure determination of membrane proteins, a number of significant practical considerations provide obstacles to the routine application of these methods, especially when it comes to resolutions ≤ 4 Å, where it becomes possible to decipher the chemical basis for the protein's function.

A. *Screening of Crystallization Trials*

The success of crystallographic methods relies on our ability to produce well-ordered crystals. The field of X-ray crystallography has made tremendous progress in developing high-throughput methods for 3D crystallization screening, using liquid-handling robots for dispensing nanoliter scale droplets across thousands of different conditions (Luft et al., 2003) and sample loading robots for placing crystals in front of synchrotron X-ray beams. These methods have been particularly important for X-ray crystallography of membrane proteins, where a shotgun approach over a huge number of variables is generally required to produce high-resolution structures (Rees, 2001). These methods are facilitated by the macroscopic nature of 3D crystals, which can be rapidly and repeatedly imaged with a light microscope. In contrast, membrane crystals are microscopic (<10 μm across and 50–500 Å thick) and therefore require electron microscopy, which necessitates multiple pipetting steps for preparing an EM grid of each sample, insertion into the microscope through an airlock, followed by evaluation at various magnifications. Crystallization requires detergent removal, typically by dialysis, and strategies for high-throughput are only beginning to be developed. Such labor-intensive procedures have severely limited the number of parameters that can be investigated in an effort to discover or to optimize crystallization conditions.

B. *Data Collection*

Once optimal conditions for crystallization and sample preparation are established, a 3D dataset includes images recorded from scores of well-ordered crystals, which generally represent a small fraction of the total number of images recorded. This inefficiency results from a high rate of electron radiation damage, which precludes prescreening of crystal quality if one wishes to record the highest resolution information in the image. This constraint makes reproducibility an essential element in the preparation of highly ordered membrane crystals. In contrast, X-ray crystallographers can screen tens to hundreds of crystals to find one that is well ordered, which can often be used to provide a complete 3D dataset.

C. *Flexibility of Membrane Crystals*

The samples used for electron crystallography comprise a single bilayer studded with a 2D array of membrane proteins (Fig. 1). The corresponding lack of physical constraints normal to this bilayer is advantageous from a physiological perspective, but means that they easily bend, curl, and sometimes break into pieces. Such defects limit the usefulness of existing structure determination software, which assumes that all molecules lie within a given 2D plane. More specifically, maximal signal-to-noise ratio is obtained when all molecules in the crystal represent a single defined orientation and thus contribute coherently to the Fourier transform. Even the slightest deviation (e.g., 1° of bending) across the crystal degrades the signal, whereas larger amounts of bending or fragmentation render the data intractable with current software (Glaeser et al., 1991).

D. *Anisotropic Resolution*

A 3D dataset is obtained by tilting crystals to a variety of angles within the electron microscope. However, there is an innate limit in the tilt angle, because crystal thickness increases dramatically above 60°. As a result, there is a missing cone of high-tilt data in the final dataset, which causes anisotropic resolution in the structure. This missing cone is exacerbated by the flexibility mentioned above, because the quality of data degrades at high tilt due to the fact that variations in the angle of view are amplified at high angle. This anisotropy in resolution causes blurring of densities perpendicular to the membrane plane, complicating the interpretation of structures and the fitting of polypeptide chains.

VIII. The Future of Electron Crystallography

We are convinced that many of these obstacles can be overcome by a concerted effort to develop the appropriate methodologies. Indeed, several developments are already underway and, if pursued to a suitable endpoint, will facilitate a higher success rate in structure determination and thus will encourage more widespread use of electron crystallography in future studies.

A. *High-Throughput Methods for Crystallization*

In a traditional manual screening, the most important parameters affecting crystallization—that is, phospholipid type, LPR, pH, temperature, detergent type, divalent cations, ionic strength, buffer, ligands, inhibitors, and additives—are surveyed in a very limited fashion. Thus, a large number of conditions must be screened in order to cover a sufficient range of relevant crystallization parameters. A key development will be to implement high-throughput screening of crystallization trials, first to establish general principles governing this process and, ultimately, to produce well-ordered membrane crystals (Fig. 4). As a start, two independent developments are underway, both involving liquid-handling robots operating on a 96-well format. The first uses a dialysis block with 50 μl sample wells and an independent 1 ml reservoir for each sample (Vink et al., 2007; Kim et al., 2010). Using a commercial liquid-handling robot to refresh reservoir solutions frequently, detergent removal over a period of 4–14 days has been demonstrated, depending on the detergent, which is comparable to results obtained in more standard dialysis setups (e.g., buttons, capillaries, or bags). The second approach to high throughput relies on the ability of cyclodextrins to effectively remove detergent from ternary mixtures of detergent, lipid, and protein and thus effect membrane protein reconstitution (Signorell et al., 2007). A custom liquid-handling robot has been designed to titrate nanoliter amounts of cyclodextrin solutions to 10–50 μl of protein samples arrayed in 96 wells (Iacovache et al., 2010). Both approaches have been effective in producing membrane crystals and have potential to screen a broad array of parameters affecting the process. Both groups have also employed liquid-handling robots to prepare negatively stained grids, using magnetic platforms to hold down Ni grids during the staining process and liquid-handling robots to carry out the pipetting steps (Coudray et al., 2010; Hu et al., 2010). These methods built on an earlier negative staining robot that employed wells drilled into a block of Teflon (Cheng et al., 2007).

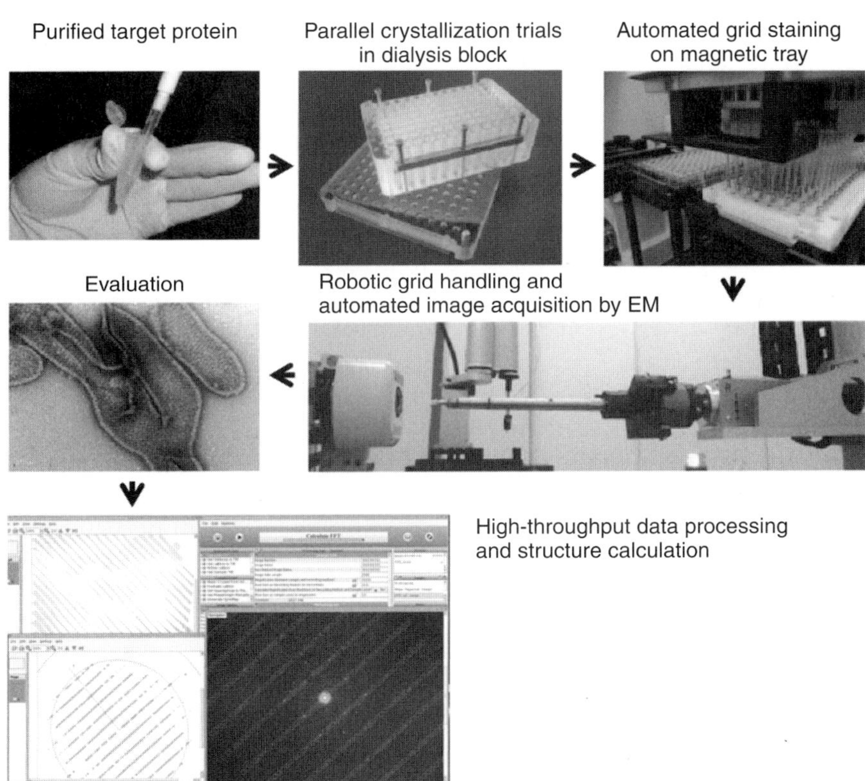

FIG. 4. Pipeline for protein structure determination by electron crystallography. Target membrane proteins are purified in detergent micelles in a stable and monodisperse form. Following the addition of lipids to form mixed micelles of protein, detergent, and lipid, the crystallization process is studied by removing dialysis in a 96-well dialysis block. The 96 crystallization conditions are harvested, transferred to EM grids, and negative stained with a liquid-handling robot. The EM grids are robotically inserted into the electron microscope, and images are recorded automatically and stored in a database. Thus, a broad range of parameters can be explored in an attempt to find large, well-ordered crystals. Finally, image processing and structure determination can be carried out using, for example, the 2dx software (Gipson et al., 2007).

B. *High-Throughput Imaging of Crystallization Trials*

These liquid-handling robots will generate large numbers of samples that must then be imaged by electron microscopy. This represents a huge bottleneck in the operation, given the logistics of operating the electron microscope. Fortunately, recent developments have resulted in several options for automated insertion and imaging of negatively stained samples. The first option involves an articulated 5-axis robotic arm that picks up individual EM grids with forceps, places them into the specimen holder, and then manipulates the holder through the airlock of a Tecnai F20 electron microscope (Potter et al., 2004). A second option divided the same procedure into two steps, employing a SCARA robot to manipulate the EM grids with a vacuum pickup and a Cartesian robot to place the holder into a JEOL 1230 electron microscope (Hu et al., 2010). In both cases, specimen insertion was controlled by the program Leginon (Potter et al., 1999), which goes on to acquire a series of representative images from each sample and to place them in a database for later evaluation. In a very different approach, a so-called auto-loader was adapted for a Tecnai T12 microscope and its grid capacity was extended by placing the commercially available 12-grid cassettes onto an 8-position carousel, such that 96 grids can be accessed by custom control software (Coudray et al., 2010). This design was reminiscent of the Gatling gun (Lefman et al., 2007), which accommodates 100 EM grids mounted in cartridges on a cylindrical drum within the vacuum of a Tecnai T12 microscope.

C. *Preparing Better Ordered Specimens*

As discussed, membrane crystals are notoriously flexible and small forces distort the crystal during adsorption to a carbon support film, which itself is not perfectly flat. Two approaches to prepare flatter specimens include the back-injection method (Kuhlbrandt and Downing, 1989) and the carbon sandwich technique (Koning et al., 2003; Gyobu et al., 2004). The carbon sandwich technique appears to be the method of choice as it consistently yields improved resolution. Grids made of Mo are frequently used due to its lower coefficient of thermal expansion and tendency to minimize crinkling of the carbon support upon freezing. Still, we believe that further developments in the design of grids and support materials will be required in the quest for high resolution. In addition, the 1.9 Å resolution obtained from AQP0 (Gonen et al., 2005) suggests that double-layered crystals have a greater stiffness than single-layered ones, and crystallization conditions could be optimized for the growth of such double-layered membrane crystals.

D. *Automated Data Collection*

Automation of low-dose imaging and collection of electron diffraction data promise to accelerate the throughput of structure determination, and current developments in software for screening crystallization trials provide a solid foundation for this work (Cheng et al., 2007; Coudray et al., 2008, 2010; Hu et al., 2010). Leginon (Potter et al., 1999) and SerialEM (Mastronarde, 2005) can automatically record low-dose images and if a strategy for locating large crystals at low magnification and evaluating crystal order through a quick peek at the electron diffraction were included, these programs would be highly effective for electron crystallography. Imaging of tilted specimens is particularly difficult, most likely due to charging phenomena. Thus, incorporating a spot-scan imaging mode, which reduces charge buildup (Downing and Glaeser, 1986), will optimize the success rate.

E. *Correction of Crystal Lattice Distortions*

Lattice distortions produced from minor lattice faults can induce substantial shifts of large coherent areas of the crystal. Moreover, crystals can grow from multiple nuclei after proteins are integrated in the bilayer thereby leading to mosaic crystals. Such distortions can be addressed by image analysis and existing methods fall into two categories: unbending (Henderson et al., 1986) and correlation averaging (CA) (Saxton and Baumeister, 1982). While both methods have been shown capable of delivering high resolution (3.5 Å)—purple membrane with unbending (Baldwin et al., 1988) and porin with CA (Sass et al., 1989)—there is a profound difference between them. Unbending strives to reconstruct a large, coherent 2D lattice which is then processed by Fourier methods, whereas CA aims to extract individual unit cells, which exhibit a high correlation with a selected reference, followed by alignment and averaging in real space. CA has the potential to accommodate rotational disorder, both in-plane and out-of-plane, and it is thus best suited for the analysis of badly fragmented or bent lattices. This approach has already shown promise in the structure determination of a secondary transporter (Koeck et al., 2007), and we believe that future developments in this area will have significant impact.

F. Robust Structure Determination Software

Improvements in data processing software are critical to the advancement of electron crystallography. Original developments in electron crystallography at the Medical Research Council (MRC) in the 1970s and 1980s produced a comprehensive set of programs that resulted in the atomic resolution structure of bacteriorhodopsin (Crowther et al., 1996). Since then, electron crystallographers have done relatively little to take advantage of the vast increase in computing power and programming infrastructure. In contrast, many refinements and new ideas have been implemented over the same time frame for single particle processing and electron tomography, developments that continue to be pursued today.

Nevertheless, several worthwhile programs have recently become available and facilitate image processing of membrane crystals. The 2dx initiative (Gipson et al., 2007) (www.2dx.unibas.ch) provides a graphical user interface to the original MRC programs and features streamlined processing solutions with optional full automation that can potentially accelerate image processing and structure determination considerably. Similarly, the XDP software handles diffraction patterns by relying on the MRC code (Hirai et al., 1999). In contrast, IPLT is a new development for processing images and electron diffraction (www.iplt.org). This program takes advantage of a modern object oriented programming architecture and incorporates new strategies for correcting lattice distortions and untangling overlapping electron diffraction patterns (Philippsen et al., 2003, 2007). IPLT is innately extensible and appears to offer a good platform for implementing new algorithms.

An emphasis on electron diffraction represents an important avenue for future development. High-resolution electron diffraction is relatively easy to collect from well-ordered membrane crystals, whereas specimen instabilities and charging effects make the process of image collection laborious and time-consuming. Thus, if molecular replacement and phase extension methods could be routinely applied to electron diffraction data, similar to what is done by X-ray crystallographers, the difficulties in obtaining high-resolution images could be bypassed and the rate of structure determination by electron crystallography could be greatly accelerated.

References

Agre, P., Preston, G. M., Smith, B. L., Jung, J. S., Raina, S., Moon, C., et al. (1993). Aquaporin CHIP: the archetypal molecular water channel. *Am. J. Physiol.* **265**, F463–F476.

Andersen, O. S., Koeppe, R. E. (2007). Bilayer thickness and membrane protein function: an energetic perspective. *Annu. Rev. Biophys. Biomol. Struct.* **36**, 107–130.

Appel, M., Hizlan, D., Vinothkumar, K. R., Ziegler, C., Kuhlbrandt, W. (2009). Conformations of NhaA, the Na+/H+ exchanger from Escherichia coli, in the pH-activated and ion-translocating states. *J. Mol. Biol.* **388**, 659–672.

Auer, M., Scarborough, G. A., Kühlbrandt, W. (1998). Three-dimensional map of the plasma membrane H+-ATPase in the open conformation. *Nature* **392**, 840–843.

Baldwin, J. M., Henderson, R., Beckman, E., Zemlin, F. (1988). Images of purple membrane at 2.8 A resolution obtained by cryo-electron microscopy. *J. Mol. Biol.* **202**, 585–591.

Beroukhim, R., Unwin, N. (1997). Distortion correction of tubular crystals: improvements in the acetylcholine receptor structure. *Ultramicroscopy* **70**, 57–81.

Berriman, J., Unwin, N. (1994). Analysis of transient structures by cryo-microscopy combined with rapid mixing of spray droplets. *Ultramicroscopy* **56**, 241–252.

Bostina, M., Mohsin, B., Kuhlbrandt, W., Collinson, I. (2005). Atomic model of the E. coli membrane-bound protein translocation complex SecYEG. *J. Mol. Biol.* **352**, 1035–1043.

Brejc, K., van Dijk, W. J., Klaassen, R. V., Schuurmans, M., van Der Oost, J., Smit, A. B., et al. (2001). Crystal structure of an ACh-binding protein reveals the ligand-binding domain of nicotinic receptors. *Nature* **411**, 269–276.

Brisson, A., Unwin, P. N. (1984). Tubular crystals of acetylcholine receptor. *J. Cell Biol.* **99**, 1202–1211.

Brugger, B., Erben, G., Sandhoff, R., Wieland, F. T., Lehmann, W. D. (1997). Quantitative analysis of biological membrane lipids at the low picomole level by nano-electrospray ionization tandem mass spectrometry. *Proc. Natl. Acad. Sci. USA* **94**, 2339–2344.

Chen, Y.-J., Pornillos, O., Lieu, S., Ma, C., Chen, A. P., Chang, G. (2007). X-ray structure of EmrE supports dual topology model. *Proc. Natl. Acad. Sci. USA* **104**, 18999–19004.

Cheng, A., Leung, A., Fellmann, D., Quispe, J., Suloway, C., Pulokas, J., et al. (2007). Towards automated screening of two-dimensional crystals. *J. Struct. Biol.* **160**, 324–331.

Cheng, A., van Hoek, A. N., Yeager, M., Verkman, A. S., Mitra, A. K. (1997). Three-dimensional organization of a human water channel. *Nature* **387**, 627–630.

Coudray, N., Beck, F., Buessler, J., Korinek, A., Karathanou, A., Remigy, H., et al. (2008). Automatic acquisition and image analysis of 2D crystals. *Micros. Today* **16**, 48–49.

Coudray, N., Hermann, G., Caujolle-Bert, D., Karathanou, A., Erne-Brand, F., Buessler, J. L., et al. (2010). Automated screening of 2D crystallization trials using transmission electron microscopy: a high-throughput tool-chain for sample preparation and microscopic analysis. *J. Struct. Biol.* [Epub ahead of print].

Crowther, R. A., Henderson, R., Smith, J. M. (1996). MRC image processing programs. *J. Struct. Biol.* **116**, 9–16.

De Feo, C., Aller, S., Siluvai, G., Blackburn, N., Unger, V. (2009). Three-dimensional structure of the human copper transporter hCTR1. *Proc. Natl. Acad. Sci. USA* **106**, 4237–4242.

Deisenhofer, J., Epp, O., Miki, K., Huber, R., Michel, H. (1984). X-ray structure analysis of a membrane protein complex. Electron density map at 3 A resolution and a model of the chromophores of the photosynthetic reaction center from Rhodopseudomonas viridis. *J. Mol. Biol.* **180**, 385–398.

Dowhan, W. (1997). Molecular basis for membrane phospholipid diversity: why are there so many lipids? *Annu. Rev. Biochem.* **66**, 199–232.

Downing, K. H., Glaeser, R. M. (1986). Improvement in high resolution image quality of radiation-sensitive specimens achieved with reduced spot size of the electron beam. *Ultramicroscopy* **20**, 269–278.

Drews, J. (2000). Drug discovery: a historical perspective. *Science* **287**, 1960–1964.

Engelman, D. M. (1971). Lipid bilayer structure in the membrane of Mycoplasma laidlawii. *J. Mol. Biol.* **58**, 153–165.

Fleishman, S. J., Harrington, S. E., Enosh, A., Halperin, D., Tate, C. G., Ben-Tal, N. (2006). Quasi-symmetry in the cryo-EM structure of EmrE provides the key to modeling its transmembrane domain. *J. Mol. Biol.* **364**, 54–67.

Gipson, B., Zeng, X., Zhang, Z., Stahlberg, H. (2007). 2dx—User-friendly image processing for 2D crystals. *J. Struct. Biol.* **157**, 64–72.

Glaeser, R. M., Zilker, A., Radermacher, M., Gaub, H. E., Hartmann, T., Baumeister, W. (1991). Interfacial energies and surface-tension forces involved in the preparation of thin, flat crystals of biological macromolecules for high-resolution electron microscopy. *J. Microsc.* **161**, 21–45.

Gonen, T., Cheng, Y., Sliz, P., Hiroaki, Y., Fujiyoshi, Y., Harrison, S. C., et al. (2005). Lipid-protein interactions in double-layered two-dimensional AQP0 crystals. *Nature* **438**, 633–638.

Gyobu, N., Tani, K., Hiroaki, Y., Kamegawa, A., Mitsuoka, K., Fujiyoshi, Y. (2004). Improved specimen preparation for cryo-electron microscopy using a symmetric carbon sandwich technique. *J. Struct. Biol.* **146**, 325–333.

Henderson, R., Baldwin, J. M., Ceska, T. A., Zemlin, F., Beckmann, E., Downing, K. H. (1990). Model for the structure of bacteriorhodopsin based on high-resolution electron cryo-microscopy. *J. Mol. Biol.* **213**, 899–929.

Henderson, R., Baldwin, J. M., Downing, K. H., Lepault, J., Zemlin, F. (1986). Structure of purple membrane from halobacterium halobium: recording, measurement and evaluation of electron micrographs at 3.5 Å resolution. *Ultramicroscopy* **19**, 147–178.

Henderson, R., Unwin, P. N. T. (1975). Three-dimensional model of purple membrane obtained from electron microscopy. *Nature* **257**, 28–32.

Hilf, R. J. C., Dutzler, R. (2008). X-ray structure of a prokaryotic pentameric ligand-gated ion channel. *Nature* **452**, 375–379.

Hirai, T., Heymann, J. A., Shi, D., Sarker, R., Maloney, P. C., Subramaniam, S. (2002). Three-dimensional structure of a bacterial oxalate transporter. *Nat. Struct. Biol.* **9**, 597–600.

Hirai, T., Murata, K., Mitsuoka, K., Kimura, Y., Fujiyoshi, Y. (1999). Trehalose embedding technique for high-resolution electron crystallography: application to structural study on bacteriorhodopsin. *J. Electron Microsc. (Tokyo)* **48**, 653–658.

Hirai, T., Subramaniam, S. (2009). Protein conformational changes in the bacteriorhodopsin photocycle: comparison of findings from electron and X-ray crystallographic analyses. *PLoS ONE* **4**, e5769.

Hiroaki, Y., Tani, K., Kamegawa, A., Gyobu, N., Nishikawa, K., Suzuki, H., et al. (2006). Implications of the aquaporin-4 structure on array formation and cell adhesion. *J. Mol. Biol.* **355**, 628–639.

Hite, R. K., Li, Z., Walz, T. (2010). Principles of membrane protein interactions with annular lipids deduced from aquaporin-0 2D crystals. *EMBO J.* **29**, 1652–1658.

Holm, P., Bhakat, P., Jegerschold, C., Gyobu, N., Mitsuoka, K., Fujiyoshi, Y., et al. (2006). Structural basis for detoxification and oxidative stress protection in membranes. *J. Mol. Biol.* **360**, 934–945.

Holm, P. J., Morgenstern, R., Hebert, H. (2002). The 3-D structure of microsomal glutathione transferase 1 at 6 A resolution as determined by electron crystallography of p22(1)2(1) crystals. *Biochim. Biophys. Acta* **1594**, 276–285.

Hong, M. (2006). Oligomeric structure, dynamics, and orientation of membrane proteins from solid-state NMR. *Structure* **14**, 1731–1740.

Hu, M., Vink, M., Kim, C., Derr, K., Koss, J., D'Amico, K., et al. (2010). Automated electron microscopy for evaluating two-dimensional crystallization of membrane proteins. *J. Struct. Biol.* **171**, 102–110.

Hunte, C., Screpanti, E., Venturi, M., Rimon, A., Padan, E., Michel, H. (2005). Structure of a Na+/H+ antiporter and insights into mechanism of action and regulation by pH. *Nature* **435**, 1197–1202.

Iacovache, I., Biasini, M., Kowal, J., Kukulski, W., Chami, M., van der Goot, F. G., et al. (2010). The 2DX robot: a membrane protein 2D crystallization Swiss Army knife. *J. Struct. Biol.* **169**, 370–378.

Jap, B. K., Walian, P. J., Gehring, K. (1991). Structural architecture of an outer membrane channel as determined by electron crystallography. *Nature* **350**, 167–170.

Jap, B. K., Zulauf, M., Scheybani, T., Hefti, A., Baumeister, W., Aebi, U., et al. (1992). 2D crystallization: from art to science. *Ultramicroscopy* **46**, 45–84.

Jegerschold, C., Pawelzik, S. C., Purhonen, P., Bhakat, P., Gheorghe, K. R., Gyobu, N., et al. (2008). Structural basis for induced formation of the inflammatory mediator prostaglandin E2. *Proc. Natl. Acad. Sci. USA* **105**, 11110–11115.

Kim, C., Vink, M., Hu, M., Love, J., Stokes, D. L., Ubarretxena-Belandia, I. (2010). An automated pipeline to screen membrane protein 2D crystallization. *J. Struct. Funct. Genomics* **11**, 155–166.

Kimura, Y., Vassylyev, D. G., Miyazawa, A., Kidera, A., Matsushima, M., Mitsuoka, K., et al. (1997). Surface of bacteriorhodopsin revealed by high-resolution electron crystallography. *Nature* **389**, 206–211.

Kistler, J., Stroud, R. M. (1981). Crystalline arrays of membrane-bound acetylcholine receptor. *Proc. Natl. Acad. Sci. USA* **78**, 3678–3682.

Koeck, P. J., Purhonen, P., Alvang, R., Grundberg, B., Hebert, H. (2007). Single particle refinement in electron crystallography: a pilot study. *J. Struct. Biol.* **160**, 344–352.

Koning, R. I., Oostergetel, G. T., Brisson, A. (2003). Preparation of flat carbon support films. *Ultramicroscopy* **94**, 183–191.

Korkhov, V. M., Tate, C. G. (2008). Electron crystallography reveals plasticity within the drug binding site of the small multidrug transporter EmrE. *J. Mol. Biol.* **377**, 1094–1103.

Krebs, A. (2003). The three-dimensional structure of bovine rhodopsin determined by electron cryomicroscopy. *J. Biol. Chem.* **278**, 50217–50225.

Kühlbrandt, W. (1992). Two-dimensional crystallization of membrane proteins. *Q. Rev. Biophys.* **25**, 1–49.

Kuhlbrandt, W., Downing, K. H. (1989). Two-dimensional structure of plant light-harvesting complex at 3.7 A [corrected] resolution by electron crystallography. *J. Mol. Biol.* **207**, 823–828.

Kühlbrandt, W., Wang, D. N., Fujiyoshi, Y. (1994). Atomic model of plant light-harvesting complex by electron crystallography. *Nature* **367**, 614–621.

Kukulski, W., Schenk, A. D., Johanson, U., Braun, T., de Groot, B. L., Fotiadis, D., et al. (2005). The 5A structure of heterologously expressed plant aquaporin SoPIP2;1. *J. Mol. Biol.* **350**, 611–616.

Kunji, E. R., von Gronau, S., Oesterhelt, D., Henderson, R. (2000). The three-dimensional structure of halorhodopsin to 5 A by electron crystallography: a new unbending procedure for two-dimensional crystals by using a global reference structure. *Proc. Natl. Acad. Sci. USA* **97**, 4637–4642.

Lee, A. G. (2004). How lipids affect the activities of integral membrane proteins. *Biochim. Biophys. Acta* **1666**, 62–87.

Lefman, J., Morrison, R., Subramaniam, S. (2007). Automated 100-position specimen loader and image acquisition system for transmission electron microscopy. *J. Struct. Biol.* **158**, 318–326.

Leifer, D., Henderson, R. (1983). Three-dimensional structure of orthorhombic purple membrane at 6.5 A resolution. *J. Mol. Biol.* **163**, 451–466.

Lindahl, E., Sansom, M. S. (2008). Membrane proteins: molecular dynamics simulations. *Curr. Opin. Struct. Biol.* **18**, 425–431.

Luft, J. R., Collins, R. J., Fehrman, N. A., Lauricella, A. M., Veatch, C. K., DeTitta, G. T. (2003). A deliberate approach to screening for initial crystallization conditions of biological macromolecules. *J. Struct. Biol.* **142**, 170–179.

Ma, C., Chang, G. (2004). Structure of the multidrug resistance efflux transporter EmrE from Escherichia coli. *Proc. Natl. Acad. Sci. USA* **101**, 2852–2857.

Mastronarde, D. N. (2005). Automated electron microscope tomography using robust prediction of specimen movements. *J. Struct. Biol.* **152**, 36–51.

Mitra, K., Ubarretxena-Belandia, I., Taguchi, T., Warren, G., Engelman, D. M. (2004). Modulation of the bilayer thickness of exocytic pathway membranes by membrane proteins rather than cholesterol. *Proc. Natl. Acad. Sci. USA* **101**, 4083–4088.

Miyazawa, A., Fujiyoshi, Y., Stowell, M., Unwin, N. (1999). Nicotinic acetylcholine receptor at 4.6 A resolution: transverse tunnels in the channel wall. *J. Mol. Biol.* **288**, 765–786.

Miyazawa, A., Fujiyoshi, Y., Unwin, N. (2003). Structure and gating mechanism of the acetylcholine receptor pore. *Nature* **423**, 949–955.

Mosser, G. (2001). Two-dimensional crystallogenesis of transmembrane proteins. *Micron* **32**, 517–540.

Murata, K., Mitsuoka, K., Hirai, T., Walz, T., Agre, P., Heymann, J. B., et al. (2000). Structural determinants of water permeation through aquaporin-1. *Nature* **407**, 599–605.

Nyholm, T. K., Ozdirekcan, S., Killian, J. A. (2007). How protein transmembrane segments sense the lipid environment. *Biochemistry* **46**, 1457–1465.

Oshima, A., Tani, K., Hiroaki, Y., Fujiyoshi, Y., Sosinsky, G. E. (2007). Three-dimensional structure of a human connexin26 gap junction channel reveals a plug in the vestibule. *Proc. Natl. Acad. Sci. USA* **104**, 10034–10039.

Philippsen, A., Schenk, A. D., Signorell, G. A., Mariani, V., Berneche, S., Engel, A. (2007). Collaborative EM image processing with the IPLT image processing library and toolbox. *J. Struct. Biol.* **157**, 28–37.

Philippsen, A., Schenk, A. D., Stahlberg, H., Engel, A. (2003). Iplt—image processing library and toolkit for the electron microscopy community. *J. Struct. Biol.* **144**, 4–12.

Pike, L. J., Han, X., Chung, K. N., Gross, R. W. (2002). Lipid rafts are enriched in arachidonic acid and plasmenylethanolamine and their composition is independent of caveolin-1 expression: a quantitative electrospray ionization/mass spectrometric analysis. *Biochemistry* **41**, 2075–2088.

Popot, J. L., Engelman, D. M. (2000). Helical membrane protein folding, stability, and evolution. *Annu. Rev. Biochem.* **69**, 881–922.

Pornillos, O., Chen, Y.-J., Chen, A. P., Chang, G. (2005). X-ray structure of the EmrE multidrug transporter in complex with a substrate. *Science* **310**, 1950–1953.

Potter, C. S., Chu, H., Frey, B., Green, C., Kisseberth, N., Madden, T. J., et al. (1999). Leginon: a system for fully automated acquisition of 1000 electron micrographs a day. *Ultramicroscopy* **77**, 153–161.

Potter, C. S., Pulokas, J., Smith, P., Suloway, C., Carragher, B. (2004). Robotic grid loading system for a transmission electron microscope. *J. Struct. Biol.* **146**, 431–440.

Preston, G. M., Carroll, T. P., Guggino, W. B., Agre, P. (1992). Appearance of water channels in Xenopus oocytes expressing red cell CHIP28 protein. *Science* **256**, 385–387.

Rees, D. C. (2001). Crystallographic analyses of hyperthermophilic proteins. *Methods Enzymol.* **334**, 423–437.

Remigy, H. W., Caujolle-Bert, D., Suda, K., Schenk, A., Chami, M., Engel, A. (2003). Membrane protein reconstitution and crystallization by controlled dilution. *FEBS Lett.* **555**, 160–169.

Rhee, K. H., Morris, E. P., Barber, J., Kuhlbrandt, W. (1998). Three-dimensional structure of the plant photosystem II reaction centre at 8 A resolution. *Nature* **396**, 283–286.

Rigaud, J. L., Mosser, G., Lacapere, J. J., Olofsson, A., Levy, D., Ranck, J. L. (1997). Bio-Beads: an efficient strategy for two-dimensional crystallization of membrane proteins. *J. Struct. Biol.* **118**, 226–235.

Ruprecht, J. J., Mielke, T., Vogel, R., Villa, C., Schertler, G. F. (2004). Electron crystallography reveals the structure of metarhodopsin I. *EMBO J.* **23**, 3609–3620.

Sass, H. J., Buldt, G., Beckmann, E., Zemlin, F., van Heel, M., Zeitler, E., et al. (1989). Densely packed beta-structure at the protein-lipid interface of porin is revealed by high-resolution cryo-electron microscopy. *J. Mol. Biol.* **209**, 171–175.

Saxton, W. O., Baumeister, W. (1982). The correlation averaging of a regularly arranged bacterial envelope protein. *J. Microsc.* **127**, 127–138.

Schenk, A. D., Werten, P. J., Scheuring, S., de Groot, B. L., Muller, S. A., Stahlberg, H., et al. (2005). The 4.5 A structure of human AQP2. *J. Mol. Biol.* **350**, 278–289.

Schertler, G. F. (2005). Structure of rhodopsin and the metarhodopsin I photointermediate. *Curr. Opin. Struct. Biol.* **15**, 408–415.

Schulz, G. E. (2000). beta-Barrel membrane proteins. *Curr. Opin. Struct. Biol.* **10**, 443–447.

Signorell, G. A., Kaufmann, T. C., Kukulski, W., Engel, A., Remigy, H. W. (2007). Controlled 2D crystallization of membrane proteins using methyl-beta-cyclodextrin. *J. Struct. Biol.* **157**, 321–328.

Stahlberg, H., Braun, T., de Groot, B., Philippsen, A., Borgnia, M. J., Agre, P., et al. (2000). The 6.9-A structure of GlpF: a basis for homology modeling of the glycerol channel from Escherichia coli. *J. Struct. Biol.* **132**, 133–141.

Stahlberg, H., Fotiadis, D., Scheuring, S., Remigy, H., Braun, T., Mitsuoka, K., et al. (2001). Two-dimensional crystals: a powerful approach to assess structure, function and dynamics of membrane proteins. *FEBS Lett.* **504**, 166–172.

Subramaniam, S., Hirai, T., Henderson, R. (2002). From structure to mechanism: electron crystallographic studies of bacteriorhodopsin. *Philos. Transact. A Math. Phys. Eng. Sci.* **360**, 859–874.

Tani, K., Mitsuma, T., Hiroaki, Y., Kamegawa, A., Nishikawa, K., Tanimura, Y., et al. (2009). Mechanism of aquaporin-4's fast and highly selective water conduction and proton exclusion. *J. Mol. Biol.* **389**, 694–706.

Tate, C. G., Ubarretxena-Belandia, I., Baldwin, J. M. (2003). Conformational changes in the multidrug transporter EmrE associated with substrate binding. *J. Mol. Biol.* **332**, 229–242.

Toyoshima, C., Unwin, N. (1990). Three-dimensional structure of the acetylcholine receptor by cryoelectron microscopy and helical image reconstruction. *J. Cell Biol.* **111**, 2623–2635.

Ubarretxena-Belandia, I., Baldwin, J. M., Schuldiner, S., Tate, C. G. (2003). Three-dimensional structure of the bacterial multidrug transporter EmrE shows it is an asymmetric homodimer. *EMBO J.* **22**, 6175–6181.

Ubarretxena-Belandia, I., Engelman, D. M. (2001). Helical membrane proteins: diversity of functions in the context of simple architecture. *Curr. Opin. Struct. Biol.* **11**, 370–376.

Ubarretxena-Belandia, I., Tate, C. G. (2004). New insights into the structure and oligomeric state of the bacterial multidrug transporter EmrE: an unusual asymmetric homo-dimer. *FEBS Lett.* **564**, 234–238.

Unger, V. M. (2001). Electron cryomicroscopy methods. *Curr. Opin. Struct. Biol.* **11**, 548–554.

Unger, V. M., Hargrave, P. A., Baldwin, J. M., Schertler, G. F. (1997). Arrangement of rhodopsin transmembrane alpha-helices. *Nature* **389**, 203–206.

Unger, V. M., Kumar, N. M., Gilula, N. B., Yeager, M. (1999). Three-dimensional structure of a recombinant gap junction membrane channel. *Science* **283**, 1176–1180.

Unwin, N. (1993). Nicotinic acetylcholine receptor at 9 A resolution. *J. Mol. Biol.* **229**, 1101–1124.

Unwin, N. (1995). Acetylcholine receptor channel imaged in the open state. *Nature* **373**, 37–43.

Unwin, N. (2005). Refined structure of the nicotinic acetylcholine receptor at 4A resolution. *J. Mol. Biol.* **346**, 967–989.

Vink, M., Derr, K., Love, J., Stokes, D. L., Ubarretxena-Belandia, I. (2007). A high-throughput strategy to screen 2D crystallization trials of membrane proteins. *J. Struct. Biol.* **160**, 295–304.

Wallin, E., von Heijne, G. (1998). Genome-wide analysis of integral membrane proteins from eubacterial, archaean, and eukaryotic organisms. *Protein Sci.* **7**, 1029–1038.

Walz, T., Hirai, T., Murata, K., Heymann, J. B., Mitsuoka, K., Fujiyoshi, Y., et al. (1997). The three-dimensional structure of aquaporin-1. *Nature* **387**, 624–627.

White, S. H. (2009). Biophysical dissection of membrane proteins. *Nature* **459**, 344–346.

White, S. H., Wimley, W. C. (1999). Membrane protein folding and stability: physical principles. *Annu. Rev. Biophys. Biomol. Struct.* **28**, 319–365.

Williams, K. A. (2000). Three-dimensional structure of the ion-coupled transport protein NhaA. *Nature* **403**, 112–115.

Xu, C., Rice, W. J., He, W., Stokes, D. L. (2002). A structural model for the catalytic cycle of Ca(2+)-ATPase. *J. Mol. Biol.* **316**, 201–211.

Yeagle, P. L. (1989). Lipid regulation of cell membrane structure and function. *FASEB J.* **3**, 1833–1842.

Zhang, P., Toyoshima, C., Yonekura, K., Green, N. M., Stokes, D. L. (1998). Structure of the calcium pump from sarcoplasmic reticulum at 8-A resolution. *Nature* **392**, 835–839.

SINGLE-PARTICLE APPLICATIONS AT INTERMEDIATE RESOLUTION

By BETTINA BÖTTCHER AND KATHARINA HIPP

School of Biological Sciences, University of Edinburgh, Edinburgh, United Kingdom

Abstract

Electron microscopy together with single-particle image processing is an excellent method for structure determination of biological assemblies that exist in multiple identical copies. Typical assemblies contain several proteins and/or nucleic acids in a defined and reproducible arrangement. Coherent averaging of electron microscopic images of 5000–100,000 copies of these assemblies allows the determination of three-dimensional structures at ca. 1–3-nm resolution. At this intermediate resolution, it is possible to map individual subunits and thus to understand the architecture and quaternary structure of the assemblies. The intermediate resolution structural information gives a solid basis on which pseudo-atomic models of the assemblies can be modeled provided that high-resolution structures of smaller entities are known. The architecture of the assemblies, their pseudo-atomic models, and knowledge on their plasticity during function give a comprehensive understanding of large-scale structural dynamics of multicopy biological complexes. In this review, we will introduce the experimental pipeline and discuss selected examples.

ADVANCES IN PROTEIN CHEMISTRY AND STRUCTURAL BIOLOGY, Vol. 81
DOI: 10.1016/S1876-1623(10)81003-X

I. Introduction

Electron microscopy and single-particle image processing at intermediate resolution (1.5–3 nm) give information on the size, shape, homogeneity, and plasticity of complexes. At this resolution, the architecture of a complex can be understood in terms of its quaternary structure and its conformational response to stimuli during catalysis, either by regulation or from the environment (e.g., pH, salt concentration). This information is important in interpreting a complex at the level of a molecular machine, where function is determined by the relative arrangement of functional modules and by their relative motions rather than at the level of chemistry that is driven by properties of specific sites.

Under perfect circumstances, three-dimensional maps (3D-maps) of a complex at intermediate resolution can be determined from a few thousand particles (ca. 5000–20,000) with a turnover of 1–4 weeks/complex. This makes the method an attractive medium-throughput structure determination approach for relatively large complexes (>200 kDa). Single-particle electron microscopy has accompanied some genome-wide complexome studies recently (Aloy et al., 2004; Han et al., 2009; Kuhner et al., 2009), where the identification of likely interaction partners in a complex entailed structural studies of that complex.

Ideally, these studies use standardized, affinity-based purification protocols for copurifying interacting proteins and identifying the individual interaction partners by mass spectrometry (Aloy et al., 2004; Kuhner et al., 2009). The results of these studies are further pruned by bioinformatic tools, which distinguish the high-confidence components from sticky proteins that are likely to be contaminants rather than true interaction partners (Gavin et al., 2002). This type of analysis predicts the plausible composition of the identified complexes. Many of the predicted complexes can be modeled by taking advantage of the structural information on protein–protein interaction sites that have been deposited in the PDB (e.g., as shown as proof of concept for the exosome; Aloy et al., 2002). However, these structural and compositional models of the predicted complexes require further experimental validation. Such experimental validation can be provided at intermediate resolution by electron microscopy and single-particle image processing, which results in an experimental 3D-map for a purified complex that should match the shape of the modeled structure of the predicted complex.

For such validation, the preferred approach is to purify the complexes with the same standardized protocol that is used for the interaction studies. However, it becomes apparent that after standard purification, only a few complexes of the complexome can be directly investigated by further structural studies (Aloy et al., 2004; Kuhner et al., 2009). This subset comprises mainly large, sturdy, and abundant complexes, which in many cases have already been structurally described in detail.

Taking this into account, the purification protocols can be adapted to select for larger and more abundant complexes (Han et al., 2009). Although this yields only a small subset of complexes of a complexome, the majority of this subset is accessible to structural studies by electron microscopy and image processing. This allows a systematic comparison of the structural properties of complexes in different organisms. One surprising outcome was the large diversity in different species of both subunit stoichiometry and quaternary structure of homologous complexes (Han et al., 2009). This suggests that the way in which different proteins interact in a complex is under considerable selective pressure during evolution.

To further explore this hypothesis, the structures of many more complexes have to be determined at intermediate resolution with high confidence in order to give reliable 3D structures for assessing and comparing quaternary structures. In order to generate these structures for a large variety of complexes and provide additional information for the elucidation of the quaternary architecture (Fig. 1), many technical challenges have to be mastered:

(1) *Sample preparation* requires specialized techniques that facilitate working with scarce, fragile, and small complexes. Respectively, these are enrichment on the electron microscopic grid, cross-linking methods, and staining. By optimizing sample preparation, a larger cross-section of the complexome becomes amenable to structural studies by electron microscopy and image processing.

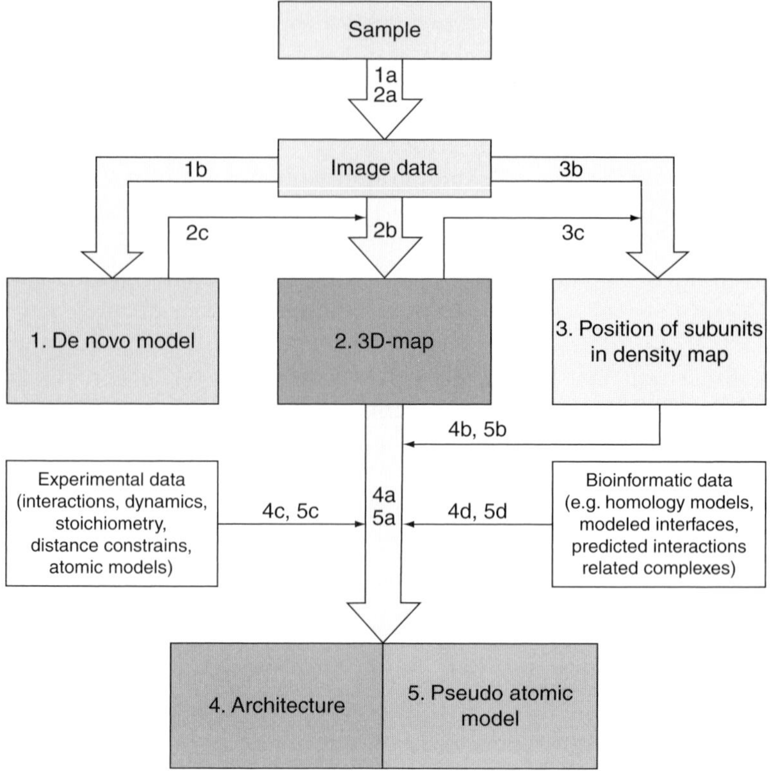

Fig. 1. Experimental pipeline for determining and interpreting structures at intermediate resolution. Electron microscopy and image processing produces (1) *de novo* models, (2) refined 3D-maps, and (3) information on the position of subunits in the 3D-map. This information is interpreted in terms of (4) the architecture of the complex and (5) pseudo-atomic models. (1) For generating *de novo* models, samples with good preservation of the low-resolution structural information are required. For this purpose, samples are often prepared by negative staining and stabilized by fixation. (1a) For data acquisition, special acquisition schemes that record tilt pairs support *de novo* model building. (1b) The image data is further processed using common lines or random/orthogonal tilt reconstruction to determine *de novo* models. The sole purpose of *de novo* models is to provide a suitable reference for starting iterative, reference-based refinement of the 3D-map. (2) For calculating intermediate resolution 3D-maps, samples are frequently prepared using cryo techniques to avoid collapse of the sample. (2a) Data acquisition aims at the generation of a larger data set with good information transfer at intermediate resolution. (2b) The maps are calculated by iterative refinement of particle orientations against reference maps. (2c) The iterative refinement is primed with a starting reference that is often provided by a *de novo* model. (3) To understand the

(2) *Image acquisition* of high-quality micrographs is a prerequisite for the efficient structure determination of complexes. Currently, sophisticated setups allow automated data acquisition on CCD or CMOS cameras without much user interaction (e.g., Leginon (Suloway et al., 2005) or Serial EM (Mastronarde, 2005)).
(3) *De novo model generation*: Most approaches in image processing rely on iterative refinement processes, which are based either on cross-common lines, or projection matching to reference projections. Generating a *de novo* model for the first reference in this iterative process is crucial for the outcome of the whole refinement.
(4) *Mapping of subunits*: To understand the quaternary structure of a complex and thus its architecture, the positions of the different proteins within the complex have to be determined. Their localization requires sophisticated labeling techniques to map sites of interest.
(5) *Building of pseudo-atomic models*: This is achieved by fitting atomic structures of smaller entities into the scaffold provided by the intermediate resolution 3D-map of the whole complex at positions determined by mapping.

The following section will discuss the different technical aspects that are important in generating reliable structural models at intermediate resolution, and will interpret them in a larger context.

---

architecture of the complex, the positions of subunits within the complex are determined. This requires specifically labeled complexes, which are usually prepared with staining techniques to provide a high signal-to-noise ratio. (3a) Data acquisition aims at small data sets that are sufficient to identify the labeled site. (3b) The positions of the labeled sites are identified in supervised classification that (3c) uses projections of the 3D-map for classification. (4) A model of the architecture of the complex is generated by (4a) the combination of the density map with (4b) the position of the subunits in the 3D-map and (4c) other experimental and (4d) bioinformatic constraints. (5) Finally, a pseudo-atomic model of the whole complex is generated. (5a) The high-resolution structures of smaller components are fitted into the intermediate resolution 3D-map of the whole complex incorporating (5b) the architecture of the complex and the position of certain sites. (5c) This pseudo-atomic model is validated and optimized against experimental constraints such as distances of known interactions or known stoichiometries. (5d) The model is further improved with bioinformatic tools that provide homology modeling and the modeling of protein–protein interactions. Reciprocally, the 3D-map together with the known position of subunits gives experimental constraints against which structural models generated with bioinformatic tools are validated.

II. Sample Preparation

Sample preparation (Fig. 1; sample) aims to stabilize the sample to prevent collapse and restructuring in the harsh environment of the electron microscope. This environment is defined by a high vacuum that leads to immediate evaporation of structural water and thus to the collapse of the sample at room temperature; high-energy electromagnetic radiation, which breaks atomic bonds, restructures the sample, and causes significant mass loss (Berriman and Leonard, 1986). Sample preparation for structure determination by single-particle electron microscopy commonly uses the negative staining technique introduced in the 1950s (Brenner and Horne, 1959) and a vitrification process developed in the 1980s (Dubochet et al., 1988).

Negative staining is an indirect method, where the structure of interest is surrounded by electron-dense material and then dried. The most common stains are uranyl acetate, uranyl formate, phosphotungstate, silicotungstate, and ammonium molybdate. Lower-density stains such as gold thioglucose (Kuhlbrandt, 1982) and methylamine vanadate (Nanovan; Hainfeld et al., 1994) have also become popular. These latter are used in special applications such as distinguishing between RNA and protein in a RNA–protein complex (gold thioglucose), and to locate small gold labels such as nanogold (Hainfeld, 1996) in a stained sample.

In negative staining, the sample leaves an imprint in the distribution of the dried stain. Because the stain is much more electron-dense than the protein, the contrast in the images is dominated by the distribution of the stain rather than by the internal structures of the protein. Negative staining is therefore an indirect preparation method with limited resolution. The resolution obtained depends on both the quality and reproducibility of the embedding and preservation of the structure of interest, and on the granular size of the stain. One problematic aspect is that the properties of the stain and the drying of the sample cause further structural damage: in particular, the loss of structural water leads to a collapse of the sample and to severe distortions perpendicular to the support film. Other artifacts can be caused by the partial embedding of the complex in stain, which gives rise to projections that represent an incomplete complex. Such distortions by collapse and incomplete embedding depend on the orientation of the complex on the sample carrier. Consequently, the projections of the structure represent differing 3D volumes depending on the spatial orientation of the object relative to the sample carrier. A full embedding of the object is achieved by the technique of sandwich negative staining, in which the sample, together with the stain, is placed between two thin layers of carbon support film (Lake et al., 1974; Golas et al., 2003). The stain distribution using this sandwich technique is highly regular. Even RNA-containing particles which tend to accumulate stain at the charged surfaces in conventional staining show an amorphous distribution of stain in the sandwich technique. Such properties of staining are compatible with structure determination at intermediate resolution, but do not circumvent the collapse of the particles.

To avoid the collapse of particles due to evaporation of structural water in the high vacuum of the electron microscope, stained samples can be frozen in liquid nitrogen or liquid ethane and imaged at temperatures below −170 °C (Golas et al., 2003), at which water does not evaporate in a high vacuum. Sandwich negative staining together with freezing is highly effective in structure preservation, and structures below a resolution of 10 Å have been reported (Golas et al., 2003). Due to the mechanism of contrast formation, these structures represent only the accessible surfaces, while internal details do not contribute significantly to the image contrast. Consequently, while structures with <10 Å resolution derived from negatively stained samples provide a very accurate representation of the outer surface which allows for precise docking of higher-resolution crystal structures, they do not necessarily resolve helices in tightly folded domains.

Despite its shortcomings, negative staining has, in recent years, become once again more popular. It requires very little material (10–200 ng/electron microscopic sample) and the preparation is quick and easy, which fits well with the demands of a medium-throughput method required by genome-wide structural complexome studies. The high contrast of stained material allows the 3D structure determination of relatively small complexes down to a molecular mass of ca. 100–200 kDa, and more easily facilitates exploration of the structural flexibility of complexes. However, taking into account the likely structural damage due to the staining and drying of the complex, the interpretation of such observed flexibility has to be handled with care. It is difficult to decide whether structural changes are functionally relevant or whether they represent a typical pattern of collapse or response to the harsh staining conditions.

In contrast to negative staining, vitrification is a direct method of preparation, where the sample is rapidly frozen in a thin layer of its own buffer. For optimal contrast, the thickness of the layer of vitrified buffer should be only slightly larger than the diameter of the particles. The contrast is formed by the difference in electron density between the vitrified buffer and the particle, which usually consists of protein and/or nucleic acids. Because the density of nucleic acids is higher than that of protein, nucleic acid-containing complexes give a higher contrast than particles of the same molecular weight that are formed of protein alone. In any case, the contrast between the vitrified buffer and the particle is considerably lower than for stained material, resulting in a much lower signal-to-noise ratio. As a consequence, complexes need to be larger for accurate determination of their spatial orientations, which in most cases means larger than ca. 500 kDa, although structures of smaller particles such as a kinase of 110 kDa (Di Luccio et al., 2007) or the small splicosomal U1-particle of 240 kDa (Stark et al., 2001) have also been determined. In both cases, the different views of the particles were very distinct, which might have aided the determination of the orientations.

Because no change in buffer conditions and no drying are involved in vitrification, it is the gentlest procedure for sample preparation that is known today. It has been demonstrated that vitrification preserves structures close to atomic resolution (Yu et al., 2008; Zhang et al., 2008a). Particles do, however, experience severe pressure if the thickness of the vitrified film is similar or smaller than their own diameter. Occasionally, fragile, viral capsids burst under these conditions although they do not show any obvious structural alterations when negatively stained.

In contrast to negative staining, sample preparation by vitrification is more time-consuming and requires more material. If particles are imaged over holes and no unspecific enrichment of the particles occurs, concentrations of ca. 6 μM are required to observe ca. 150 particles in a field of view with a diameter of 1 μm and a film thickness of 50 nm. For a 500 kDa complex, this is comparable to a concentration of 3 mg/ml and 6 μg of protein per sample. However, only a few complexes behave ideally, and usually an unspecific enrichment occurs even on the air–water interface. With the need to carry out structure determination with less and less material, enrichment seems to be an attractive concept. Many complexes enrich on thin carbon support films by a factor of 2–10 compared to preparations over holes without support film. For larger complexes, the noise added by the support film does not significantly interfere with the image processing. The carbon film also reduces beam-induced movement and charging, thus increasing the yield of high-resolution micrographs. Consequently, additional support films are now widely used for the preparation of vitrified samples. Furthermore, demands to work with even lower concentrations fostered the development of the affinity grid (Kelly et al., 2008), which is based on Ni-NTA lipids that are immobilized on a carbon support film. These lipids interact specifically with His-tagged proteins and enrich the tagged complex on the support film. This not only allows for lower concentrations in sample preparation but also enables so-called on-grid purification, where the His-tagged complex is directly pulled out of cell lysate with the affinity grid, using one-step affinity purification. The affinity grid is compatible with both vitrification and negative staining, and thus offers an ideal tool for sample preparation in exhaustive complexome studies. However, one of the major drawbacks of the method is that the use of tags restricts the spatial orientations that the particle can adopt. For image reconstruction, which relies on a good spread of orientations, this is a serious limitation, which can be partly overcome by tagging different subunits of the same complex.

Some complexes are very labile and do not withstand interactions with surfaces or stains. They disintegrate during absorption onto the sample carrier, during staining, and even during purification by gel filtration. Disintegration makes even low-resolution structure determination difficult, if not impossible. There are various reasons for the instability: the intrinsic lability of some of the complexes, the propensity of others to be destabilized by tags, proteolysis, dilution, or unfavorable buffer conditions. In general, the stability of such labile complexes can be improved by fixation with glutaraldehyde. However, simply adding glutaraldehyde often leads to irregular aggregation of the protein, which does not allow for any further structure determination. To solve this problem, an optimized fixation protocol for structure determination has been developed, known as Grafix (Kastner et al., 2008). The basic idea is that the complex of interest is fixed on a glycerol gradient during ultracentrifugation. While the complex is migrating through the gradient and stabilized by the high gravity, it is exposed to increasing glutaraldehyde concentrations ranging from 0% at the top of the gradient to 0.15% at the bottom. By coupling fixation to a density centrifugation step, an efficient separation of fragments of different sizes is achieved, producing homogenous complexes in the different fractions. Although doubts have been raised that fixation is compatible with high-resolution structural determination, tests with the ribosome have confirmed that fixation does not alter the structure, and that 15 Å resolution is achievable (Kastner et al., 2008). The Grafix protocol is also compatible with both negative staining and vitrification, giving access to the full range of possible sample preparation methods.

Grafix enabled the structure determination of many complexes that were otherwise too heterogeneous and fragile for structural investigations. A recent remarkable example is the 100S ribosome in hibernation, where Grafix greatly increased the number of dimeric 100S ribosomes compared to monomeric 70S ribosomes (Kato et al., 2010).

Grafix also became an important tool for complexome studies. These studies often use tandem affinity purification (TAP; Puig et al., 2001), where two affinity tags of different functionality are fused to the bait protein for copurification of interacting proteins. After the second stage of TAP, the purified complexes are often too heterogeneous and too dilute for further structure determination. Here, Grafix can efficiently replace the second stage of affinity purification. Dot-blot assays against the second tag identify the fractions that contain the bait protein and thus the complex of interest. This gives an efficient method for the preparation of complexes for electron microscopy in medium-throughput complexome studies, where the amount of material is limited and the purification protocols are not optimized for the individual complexes (Kuhner et al., 2009).

III. Data Acquisition

Structure determination at intermediate resolution does not pose any special challenges for the equipment. In principle, a conventional microscope with a thermionic gun is sufficient to collect data (Fig. 1; image data) between 1.5- and 3-nm resolution; with such equipment, image reconstructions at 1-nm resolution have been reported for icosahedral capsids (Trus et al., 1997; Kronenberg et al., 2001). However, a field emission gun, having better spatial coherence, is more suitable for the structure determination of unstained, vitrified complexes where phase contrast is mainly generated by defocusing. The limited spatial coherence of the electron source causes a defocus-dependent envelope function. For thermionic emission, typical defocus values of 2–3 μm lead to a significant loss in contrast transfer at resolutions as low as 1.5 nm (Wade, 1992).

Another requirement is for the equipment to be able to collect data under low dose conditions in order to minimize beam damage. This is true even for negatively stained samples at room temperature which at first glance appear to be quite stable. This is due to the fact that irradiation with electrons causes major rearrangements in the distribution of the stain (Unwin, 1974), interfering with structure determination even at intermediate resolution.

The use of CCD or CMOS cameras with at least 4k × 4k pixels speeds up data acquisition. Such electronic cameras greatly reduce the amount of time required between data acquisition and image processing, making the time-consuming tasks of film processing and scanning obsolete. An additional advantage is the better contrast transfer of the CCD cameras at low spatial frequencies (Sander et al., 2005), which facilitates the determination of the orientations of particles and *de novo* model building.

Further gains to be made by using an automated setup are few, since there is a considerable overhead involved in setting up the system. Considering the relatively small number of particles that are required for structure determination at intermediate resolution, fully automated data acquisition only becomes important if data for many complexes is to be recorded under similar conditions in a short period of time.

A further consideration is the generation of *de novo* models (Fig. 1). Options include random conical tilt reconstruction (Radermacher, 1988) and orthogonal tilt reconstruction (Leschziner and Nogales, 2006), both of which require tilted pairs of the same area. The acquisition of this kind of data is greatly facilitated by semiautomation that supports accurate image tracking during tilting (Mastronarde, 2005; Yoshioka et al., 2007).

IV. *De Novo* Model Building

Structure determination of new complexes relies upon *de novo* model-building from electron-microscopic images (Fig.1; 1b). The *de novo* models are used as starting references (Fig. 1; 2c) for further reference-based iterative refinement. The determination and choice of suitable *de novo* models for refinement is the most critical step for further structure determination. To generate *de novo* models, either common line methods (Crowther et al., 1970; van Heel, 1987), random conical tilt reconstruction (Radermacher, 1988), or the related orthogonal tilt reconstruction (Leschziner and Nogales, 2006) are most frequently used.

The common lines method allows the determination of the relative orientations between different projections of a unique object by comparison of the projections against one another (Fig. 1; 1b). No special experimental setup is needed for the data acquisition. However, it is important that the particles assume random orientations and vary in their relative rotations around more than one virtual tilt axis. Furthermore, all projections that are included in the analysis should be consistent with a single volume. If the latter condition is fulfilled, the Fourier transforms of the projections are central sections (same origin) through the Fourier transform of the single volume. The sections vary in their relative orientations according to the relative orientations of the imaged object (DeRosier and Klug, 1968; Crowther et al., 1970). As a consequence, the different sections intersect with each other. At the lines of intersection (common lines), the information in both sections is the same. The position of the common lines in each section depends on the relative orientations of these sections and thus on the relative orientations of the imaged objects in real space. Three different, correctly assigned projections determine the orientation of the object in space. However, the absolute handedness of the object remains unknown, because no absolute, external reference point is used for determining the relative orientations.

The exact determination of the relative orientations of the first three projections is most important in assigning the relative orientations of the remaining projections. For the first three projections, only two lines in each Fourier section fix the relative orientations in an asymmetric object. These make up only a small proportion of the available image information for each section/projection, and the reliable assignment of orientations with such little information can be error-prone. Consequently, for successful *de novo* model building, a few requirements should be met: (1) A high signal-to-noise ratio in the first three projections is needed to overcome problems caused by the small fraction of image information that is used for the determination of orientations. The signal-to-noise ratio of individual particle projections is usually too low, and therefore, averages of similar projections are used instead. These similar projections are identified by statistical methods such as multivariate statistical analysis (MSA; van Heel and Frank, 1981; van Heel, 1984) or maximum likelihood methods (Scheres et al., 2007). (2) For good sampling of the first three Fourier sections by common lines, the relative orientations in real space need to differ by relative rotations around different virtual rotation axes, otherwise, the common lines are colinear. The best distribution of common lines in the Fourier sections is achieved with three orthogonal views. (3) The chosen projections need to represent the same 3D object. Although this appears trivial, in heterogeneous particle populations, it is often difficult to distinguish between those projections that represent the same object in different orientations and those that represent slightly different objects.

For random conical tilt reconstructions, two different views of the same object are required. This is achieved by recording pairs of micrographs where one micrograph is taken from a highly tilted (typically 50–60°) sample and the other one from an untilted sample. Particle projections are selected pair-wise from the micrographs of tilted and untilted objects. The projections of the untilted particle are aligned in plane in respect to each other and then classified according to their similarity (see above). 3D volumes are then calculated from the related images of tilted particles of the same class. The relative spatial orientations of the related, tilted particles are given by the in-plane rotation of the aligned, untilted particle projections, the orientation of the tilt axis, and the tilt angle. These parameters can be determined with high precision, giving a reliable volume for each class. Unlike the common lines method, the absolute handedness of the volume is also known, because the tilt angle between related tilted and untilted particles is known.

One of the major drawbacks of the random conical tilt method is the limited tilt angle, which inhibits acquisition of certain volume information in the *z*-direction. To recover this information, volumes from classes representing different views of the object are aligned with respect to each other and averaged. These aligned volumes can also be tested by 3D classification, which identifies the conformational plasticity of the object (Sander et al., 2010) and generates volumes for the different conformations.

Alternatively, the missing information in the *z*-direction can be recovered by using a related tilt reconstruction scheme—orthogonal tilt reconstruction. Here, the object is tilted by $+45°$ and $-45°$. Similar to the random conical tilt reconstruction, one of the micrographs is used for the in-plane alignment of particles and their classification, and the other for reconstructing the volume in each particular class. Due to the tilt geometry, there is a 90-tilt angle between related tilted particles. Consequently, there is no missing information in *z*-direction, and given that there are enough different views in a class, volumes have isotropic resolution.

Random conical tilt and orthogonal tilt reconstructions are very robust methods for building *de novo* models (Fig. 1; 1b). However, certain practical limitations make them difficult to use. Both require the acquisition of micrographs from highly tilted samples. For cooled, vitrified samples, this can be challenging, since micrographs often have a much lower resolution perpendicular to the tilt axis due to charging. In addition, the widely used side entry cryo-holders drift considerably after tilting, which makes data acquisition slow and further reduces the image quality. Another technical obstacle is the intrinsically low signal-to-noise ratio of unstained particles which, especially for smaller complexes, makes it difficult to identify matching pairs of particles across both micrographs. For these practical reasons, many tilt reconstructions are based on negatively stained particles at ambient temperature rather than on vitrified samples. However, under these conditions, particles often suffer from directional flattening perpendicular to the plane of the sample carrier. The resulting volumes represent differently flattened objects, which are inconsistent with a unique volume viewed from different directions. Despite these limitations, however, the resulting volumes are still useful for establishing the absolute handedness of the reconstruction and for recognizing general common features in the differently distorted objects. Nevertheless, the use of the volumes as references in iterative alignment has to be handled with care and needs a critical evaluation of the results.

De novo model building can be time-consuming and error-prone regardless of whether the common line or tilt reconstruction methods are used. One general strategy is to test different approaches and to follow the convergence of the resulting volumes to a plausible solution, a process now assisted by web-based wrappers (Voss et al., 2010) which interface to different methods. With the ambiguities inherent in the determination of *de novo* models, there is high demand for criteria that allow the identification of useful reference models. Unfortunately, up until now there has been no objective measure which can be used to determine whether a reference model is correct or useful. However, there are certain properties that a good reference model should have: (1) Projections of the model should represent most of the characteristic views that were identified in the classification, before the data was biased by alignment to produce projections of the model. (2) If the initial data shows no preferential orientations, the particle orientations determined by projection matching or cross-common lines against the reference model in iterative refinement should also show no preferential orientation. (3) Projections of the model (masked with a tight mask or projected with a threshold that generates a volume of sensible size) and characteristic views assigned with the same orientation should be similar in all main features. This should be true for all characteristic projections.

The fulfillment of these basic criteria is not necessarily sufficient to guarantee a correct model, but the failure to satisfy them makes it highly unlikely that the determined model is a good representation of the experimental image data.

V. Mapping of Subunits

Image reconstructions of complexes at intermediate resolution can be used as a scaffold for the interpretation of other structural and biophysical data (Fig. 1; 4c and 5c). It is therefore important to know where key functional elements are located in the complex. In many cases, this is done simply by placing high-resolution crystal structures of individual proteins at appropriate positions in the 3D scaffold (Fig. 1; 5a, 5c, 5d). However, at intermediate resolution, this can be ambiguous especially if the absolute handedness of the complex is also not known (e.g., by starting with a model determined by common lines alone). Hence for complexes with unknown architecture, experimental localization of individual proteins in the complex is necessary (Fig. 1; 3 position of subunits) to provide anchor points which help to fit high-resolution structures of subunits into the map and therefore to validate the assignment.

In an ideal situation, the label should be directly and unambiguously visible in the raw data to circumvent time-consuming image processing procedures, and should not produce any unspecific background. A label should be characterized by high affinity and specificity. In addition, it is desirable to have a label that is adjustable in size to facilitate recognition of the label on differently sized complexes. In reality, the ideal, universally usable label has probably not been discovered yet, but there are several methods currently in use, each of which has its own advantages and disadvantages.

Using nanogold and related organometallic clusters has the advantage of being a label that can be directly observed on the specimen. In the early

1990s, Milligan et al. used monomaleimide nanogold to localize a cysteine residue in a helical assembly (Milligan and Safer, 1990). Later, Wilkens and Capaldi (1992) used nanogold for the *Escherichia coli* F_1 ATPase to label subunits, employing either a cysteine already present in a subunit or after introducing cysteine residues by site-directed mutagenesis.

The C terminus of the assembly domain of the hepatitis B virus capsid protein has been localized by image reconstruction of undecagold-labeled capsids (Zlotnick et al., 1997). The monomaleimidyl undecagold, which contains an even smaller gold cluster of 8 Å compared to 14 Å for nanogold, was coupled to a mutant of the capsid protein containing a single cysteine at its C terminus. It was possible to unequivocally assign undecagold clusters to additional densities under the fivefold and quasi-sixfold vertices in the 3D reconstruction of viral capsids imaged under cryo conditions even though less than 20% of the protein was labeled. This labeling efficiency is a result of the fact that only one subunit per vertex was modified with the label which in turn sterically prevents other labeled subunits from assembling at the same vertex.

Whereas nanogold is directly visible in the specimen, the smaller undecagold is only barely detectable in unprocessed data of vitrified samples. Using an even smaller cluster would have the advantage of potentially less steric hindrance so that more positions could be reached with the label. This has been demonstrated in the tetrairidium cluster linked via maleimide to the C terminus of the assembly domain of the hepatitis B virus capsid protein (Cheng et al., 1999). Ir_4 is small enough to label the C terminus at the interior of intact capsids, something which is not possible with undecagold, and although this cluster is not directly visible in cryo micrographs, it is clearly detectable in the 3D density maps.

In all the examples described above, it is a prerequisite that the attachment site for the organometallic cluster is surface-accessible, a condition which restricts the number of possible residues that can be used for labeling. In cases where there are several, surface-accessible cysteines in the complex, these need to be replaced to generate unambiguous results. This dramatically increases the work required to generate mutants that allow specific labeling of unique sites. Another disadvantage of gold is its tendency to nonspecific labeling as well as to aggregation of the label over time.

Wolf et al. (2009) developed quite an elaborate approach to localize the intron and the 5′ and 3′ exons of a model pre-mRNA within the spliceosomal B complex. They used antibodies as well as antibodies with attached colloidal gold-coated protein A to facilitate detection of the bound antibody. The model pre-mRNA was tagged with MS2 aptamers, and purified spliceosomal complexes were incubated with a fusion protein of MS2 coat protein and maltose binding protein (MBP). The RNA within the complex was then detected either directly with an antibody targeting the MBP moiety or after stabilizing the complex with the bound antibody by Grafix and adding the colloidal gold coated with protein A.

Instead of using an antibody coupled to gold for directly visualizing a specific site within the complex, the antibody itself can be utilized as an additional density pointing to the subunit of interest. Several ribosomal proteins and preribosomal factors have been localized on the Rix1 particle with the help of an antibody targeting a small peptide tag fused to the respective subunit (Ulbrich et al., 2009). Because of their size, smaller peptide tags like the HA tag in the aforementioned analysis should interfere less with complex stability compared to larger tags or whole domains fused to subunits of the complex. The labeling efficiency of subunits with monoclonal antibodies is far below 100% which means that labeled and unlabeled complexes have to be separated most often by computational means. In addition, this labeling approach requires a purification step to remove unbound antibodies.

During recent years, the green fluorescent protein (GFP) has revolutionized light microscopy—GFP fused to cellular proteins is a good marker for localizing protein within a cell. Recently, GFP has also been used as a tool to map subunits within a complex in electron microscopy. Compared to small peptide tags, GFP is large enough to be detected as an additional density that does not necessarily require further enlargement by an antibody. The motor domain of the cytoplasmic dynein heavy chain is a successful case in which only GFP/BFP was used for the localization of AAA+ modules as well as the N- and C terminus of the motor domain in 2D class averages of negatively stained samples (Roberts et al., 2009). In a similar approach, visualization of GFP fused to core septins established that the heterotetrameric core complex is symmetric (John et al., 2007). Here again, GFP was detected as an extra globular density in 2D class averages of negatively stained specimens. Direct localization of GFP in a cryo-EM map requires a rigid and confined orientation of the GFP, as has been achieved with some icosahedral capsids (Charpilienne et al., 2001; Conway et al., 2010).

The advantage of adding a whole domain or protein tag to a subunit of the complex is that the modification is done genetically, resulting in a stoichiometric labeling with no additional purification step required. However, fusing a domain/protein to the target protein may severely disturb the complex stability (Böttcher et al., 2006), possibly bringing about particle heterogeneity and, in the worst case, disintegration of the complex.

Ideally, one would like to combine the advantages of having an electron-dense metal cluster that is small and directly visible with cloning. A promising candidate for this purpose is metallothionein, which is a 60 amino acid long protein tag that can load ca. 12–20 gold atoms (Mercogliano and DeRosier, 2007). By fusing several tags into a row, the size of the gold cluster can be controlled, offering the possibility of differential labeling.

The most recent development in the mapping of subunits combines a small, clonable tag fused to the subunit of interest and a second component that produces a marker upon addition to the purified complex containing the tag. This molecular pointer is large and rigid enough to be directly visible in the raw images of the complex. One example is the use of Spire, a protein that nucleates assembly of an actin filament (Stroupe et al., 2009). The sequence for a truncated version of Spire that is sufficient to initiate filament growth is cloned to the desired subunit, and the expressed and purified labeled complex is then mixed with actin monomers. Polymerization of actin is halted by adding cytochalasin B at the desired point, adjusting the length of the pointer by quenching filament growth. Another molecular pointer is generated by the interaction between the yeast dynein light chain (Dyn2) and a dynein light chain-interacting domain (DID) of a nucleoporin (Flemming et al., 2010). Dyn2 homodimers form globular domains that are aligned between two DID strands giving rise to a rod-like structure (Fig. 2). One DID domain is used to label the subunit of interest, and during purification, Dyn2 is added along with a second Flag-tagged DID domain so that the pointer is formed.

The advantage of these new labeling methods is that a relatively small tag (around 10 kDa) that is less likely to interfere with complex stability is genetically fused to the subunit of interest and is therefore present stoichiometrically in the complex. The actual pointer to the subunit is much larger (~130 kDa for the DID-Dyn2 label, varying for the actin label

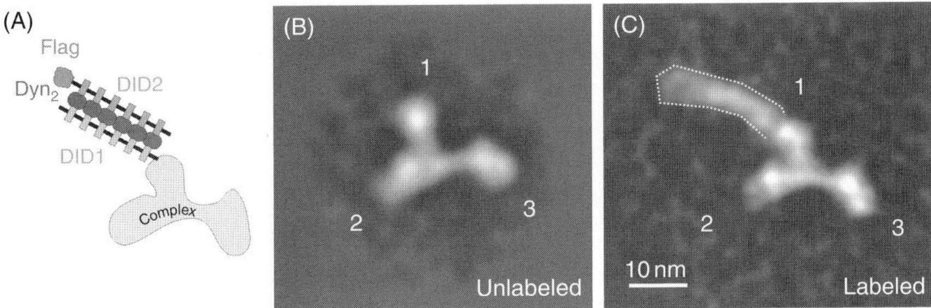

Fig. 2. Molecular pointer—DID1-Dyn2-DID2 (adapted from Flemming et al., 2010). (A) Labeling scheme: the site of interest in the complex (gray) is tagged with a dynein light chain-interacting domain from Pac11 (DID1, green). This domain binds to dynein light chain dimers (Dyn2, blue) which form globular domains and to a second dynein light chain-interacting domain from Nup 159 (DID2, green) which is coupled with a Flag tag (red). The Flag tag enables purification of the labeled complex. The fully assembled label resembles pearls on a string. (B) The unlabeled Nup-84 complex consists of Nup120-Nup145C-Nup85-Seh1-Sec13 and forms three prominent domains (indicated by numbers). (C) Seh1 of the Nup84 complex is labeled with the DID1-Dyn2-DID2 label, which forms a prominent molecular pointer (outlined by a white, dotted line). (See color plate 3).

depending on the filament length) and directly visible in the raw images. The practicability and versatility of these labels have still to be explored.

Information obtained by localizing a subunit within a complex provides a deeper insight into the architecture of the complex and may facilitate the fitting of high-resolution structures into the density map (Fig. 1; 4b and 5b).

VI. Building of Pseudo-Atomic Models

For some complexes, high-resolution structures of smaller components (subunits or subcomplexes) are available. These structures are either derived from homology modeling or from structure determination by NMR or X-ray crystallography, and can be fitted into the EM scaffold at intermediate resolution to generate a pseudo-atomic model of the whole complex (Fig. 1, 5a–5d). This strategy was first used in the early 1990s for viral capsids and the identification of receptor or Fab-binding sites (Wang et al., 1992; Olson et al., 1993; Stewart et al., 1993). While at that time it was still controversial as a scientifically acceptable technique, it is now in general use. The resulting hybrid models can be interpreted and probed in much greater detail than the EM volume or the individual high-resolution structures alone.

Depending on the shape of the complex, it is possible to place the high-resolution structures of the smaller entities with high accuracy (e.g., Fig. 3). This is especially true if most of the surface of the smaller component overlaps with the surface of the complex, though it becomes more difficult if the smaller component is buried and does not share surface-exposed sites with the complex. The fitting of the smaller fragments is assisted by measuring local correlation supported by programs

Fig. 3. Pseudo-atomic model of a complex (Diepholz et al., 2008). For many of the subunits and subcomplexes of the V-ATPase, high-resolution structural models exist or can be generated by homology modeling. Placing these models (ribbon diagrams) into the 3D-map of the whole complex (gray) generates a pseudo-atomic model for most of the V-ATPase. For some subunits, no high-resolution structural information exists. Their position and shape can be deduced by comparison of the density map with the pseudo-atomic model (e.g., subunit D, solid purple density; subunit a, solid blue density). (See color plate 4).

such as Situs (Wriggers and Birmanns, 2001) or Chimera (Pettersen et al., 2004), or by optimizing the surface overlap between the smaller entity and the whole complex, as in 3som (Ceulemans and Russell, 2004).

Often, discrepancies occur between the shape of the whole complex and the smaller entities. These discrepancies may either reflect structural rearrangements of the smaller entities in the context of the complex or simply be errors in the intermediate resolution maps. In many cases, further experiments by other methods are required to distinguish between errors in the map and functionally relevant conformational changes. If the experimental validation confirms that rearrangements of the smaller components are likely to occur, further analysis by flexible fitting can be used to adapt the shape of the smaller entity to match the observed volume. This flexible fitting makes use either of mechanical network analysis as used in motion tracking (e.g., Situs Wriggers and Birmanns, 2001) or of normal-mode analysis (Tama et al., 2004; Suhre et al., 2006) which probes the low-frequency modes of elastic networks to identify the likely large-domain movements of the smaller entity. The conformation of the smaller component is adapted to optimally match the intermediate resolution volume of the whole complex. Such an analysis highlights where flexible and tense regions occur within the complex and which parts can potentially change their positions by large distances, both factors which are important in understanding the mechanics of a complex.

For the building of pseudo-atomic models into intermediate resolution scaffolds, it is useful to have as many constraints as possible. Some of these constraints are derived from the localization of individual subunits by electron microscopy and image processing of specifically labeled complexes (see above), while other constraints might be distance constraints between certain positions of the complex. Such distances can be provided by fluorescence resonance energy transfer (FRET) measurements of labeled complexes or by cross-linking and mass spectrometry (Chen et al., 2010). The latter is especially attractive, because the constraints are accurate enough to pinpoint likely interaction sites. For large complexes like the RNA polymerase II, around 100 high-confidence interactions can be identified (Chen et al., 2010) and used to scrutinize the newly generated pseudo-atomic models.

VII. Integrating Structural Data at Intermediate Resolution with Other Biophysical Information

A generally accepted pipeline for generating reliable 3D scaffolds at intermediate resolution has now been established (Fig. 1; 1a–2c), which generates a fast-increasing output of intermediate resolution structures. Nowadays, these structures together with subunit localization and model-building are central in deducing the architecture of novel complexes and in understanding the building principles of complexes in general. However, these structures also have a major impact on cell and systems biology. For example, one of the important questions concerns how complexes are distributed in cells, in what respect they form interacting networks, and how these structures are maintained and changed under different physiological conditions. Such questions are often approached using tomography for large complexes ($\gg$500 kDa). Known structures of complexes are used as references in template matching algorithms to identify and localize the complexes in the cell (or sections of cells; Böhm et al., 2000; Beck et al., 2009). The template matching requires appropriate templates, which resemble the complexes to be matched as closely as possible. With the observed variability in the quaternary structure across species (Han et al., 2009), it is insufficient to generate templates for these complexes solely based on homology modeling. Here, experimentally determined intermediate resolution, high-quality structural models of the complexes of the respective organisms will be more reliable templates and thus generate more confident matches.

Another important aspect of complexes is structural dynamics. They breathe, move, flex, and twist during their activity. Part of this plasticity is already recognizable at intermediate resolution and can be identified and sorted by statistical methods (see Chapter 4). An impressive example of this is recent work on the translating ribosome (Fischer et al., 2010), in which more than two Mio particles have been analyzed and eight distinct conformational states have been identified using electron cryo-microscopy and single-particle image analysis. Considering the logical order of translation and the similarity between the different conformations, the observed conformational states could be ordered to reflect the progression of events in translation. However, such electron microscopic analysis does not include the absolute timescale, and the sequence of the conformational changes remains speculative. These dynamic aspects are better addressed by spectroscopic methods, such as single-molecule FRET measurements, which are accurate in the temporal scale but lack spatial resolution. Combining the detailed structural information on the different conformations of a complex with spectroscopic data allows identification of the sequence in which the different states occur during function and the assignment of an absolute timescale to the different steps. For the ribosome, the wealth of structural and single-molecule spectroscopic information already provides a comprehensive understanding of the intermediate steps during translocation and an interpretation of these processes on the thermodynamic level (see Munro et al., 2009 for review).

The fitting of structures of individual crystallized subunits and subcomplexes into the intermediate resolution EM scaffold of the whole complex often shows severe discrepancies. This suggests rearrangement and tension in the complex, which can have significant implications for the function. One example is the V-ATPase, the structure of which has been determined at intermediate resolution (Diepholz et al., 2008; Zhang et al., 2008b; Muench et al., 2009) and compared with the crystal structures (Sagermann et al., 2001) and small angle X-ray scattering solution structures of smaller entities (Diepholz et al., 2008). The discrepancies suggest tensions within the complex, which map to subunits that are pivotal for controlling regulatory disassembly. Such insights arising from comparison of the stable structures of smaller components with the structures of the whole complex provide valuable information in helping to understand complexes as molecular machines from a mechanistic point of view. This opens up a whole new perspective that bridges the information gap between biology and nanotechnology.

Taken together, the main importance of electron microscopic structures at intermediate resolution is the understanding of a complex's architecture and its large-scale plasticity. This requires a comprehensive experimental setup, of which electron microscopic structure determination by single-particle methods is an important part, and one that is tightly bound to other disciplines such as spectroscopy, proteomics, bioinformatics, and X-ray crystallography.

References

Aloy, P., Bottcher, B., Ceulemans, H., Leutwein, C., Mellwig, C., Fischer, S., et al. (2004). Structure-based assembly of protein complexes in yeast. *Science* **303**, 2026–2029.

Aloy, P., Ciccarelli, F. D., Leutwein, C., Gavin, A. C., Superti-Furga, G., Bork, P., et al. (2002). A complex prediction: three-dimensional model of the yeast exosome. *EMBO Rep.* **3**, 628–635.

Beck, M., Malmstrom, J. A., Lange, V., Schmidt, A., Deutsch, E. W., Aebersold, R. (2009). Visual proteomics of the human pathogen Leptospira interrogans. *Nat. Methods* **6**, 817–823.

Berriman, J., Leonard, K. R. (1986). Methods for specimen thickness determination in electron microscopy. II. Changes in thickness with dose. *Ultramicroscopy* **19**, 349–366.

Böhm, J., Frangakis, A. S., Hegerl, R., Nickell, S., Typke, D., Baumeister, W. (2000). From the cover: toward detecting and identifying macromolecules in a cellular context: template matching applied to electron tomograms. *Proc. Natl. Acad. Sci. USA* **97**, 14245–14250.

Böttcher, B., Vogel, M., Ploss, M., Nassal, M. (2006). High plasticity of the hepatitis B virus capsid revealed by conformational stress. *J. Mol. Biol.* **356**, 812–822.

Brenner, S., Horne, R. W. (1959). A negative staining method for high resolution electron microscopy of viruses. *Biochim. Biophys. Acta* **34**, 103–110.

Ceulemans, H., Russell, R. B. (2004). Fast fitting of atomic structures to low-resolution electron density maps by surface overlap maximization. *J. Mol. Biol.* **338**, 783–793.

Charpilienne, A., Nejmeddine, M., Berois, M., Parez, N., Neumann, E., Hewat, E., et al. (2001). Individual rotavirus-like particles containing 120 molecules of fluorescent protein are visible in living cells. *J. Biol. Chem.* **276**, 29361–29367.

Chen, Z. A., Jawhari, A., Fischer, L., Buchen, C., Tahir, S., Kamenski, T., et al. (2010). Architecture of the RNA polymerase II-TFIIF complex revealed by cross-linking and mass spectrometry. *EMBO J.* **29**, 717–726.

Cheng, N., Conway, J. F., Watts, N. R., Hainfeld, J. F., Joshi, V., Powell, R. D., et al. (1999). Tetrairidium, a four-atom cluster, is readily visible as a density label in three-dimensional cryo-EM maps of proteins at 10-25 A resolution. *J. Struct. Biol.* **127**, 169–176.

Conway, J. F., Cockrell, S. K., Copeland, A. M., Newcomb, W. W., Brown, J. C., Homa, F. L. (2010). Labeling and localization of the herpes simplex virus capsid protein UL25 and its interaction with the two triplexes closest to the penton. *J. Mol. Biol.* **397**, 575–586.

Crowther, R. A., DeRosier, D. J., Klug, A. (1970). The reconstruction of a three-dimensional structure from projections and its application to electron microscopy. *Proc. R. Soc Lond. A* **A317**, 319–340.

DeRosier, D. J., Klug, A. (1968). Reconstruction of three dimensinal structures from electron micrographes. *Nature* **217**, 130–134.

Di Luccio, E., Petschacher, B., Voegtli, J., Chou, H. T., Stahlberg, H., Nidetzky, B., et al. (2007). Structural and kinetic studies of induced fit in xylulose kinase from Escherichia coli. *J. Mol. Biol.* **365**, 783–798.

Diepholz, M., Venzke, D., Prinz, S., Batisse, C., Florchinger, B., Rossle, M., et al. (2008). A different conformation for EGC stator subcomplex in solution and in the assembled yeast V-ATPase: possible implications for regulatory disassembly. *Structure* **16**, 1789–1798.

Dubochet, J., Adrian, M., Chang, J. J., Homo, J. C., Lepault, J., McDowall, A. W., et al. (1988). Cryo-electron microscopy of vitrified specimens. *Q. Rev. Biophys.* **21**, 129–228.

Fischer, N., Konevega, A. L., Wintermeyer, W., Rodnina, M. V., Stark, H. (2010). Ribosome dynamics and tRNA movement by time-resolved electron cryomicroscopy. *Nature* **466**, 329–333.

Flemming, D., Thierbach, K., Stelter, P., Bottcher, B., Hurt, E. (2010). Precise mapping of subunits in multiprotein complexes by a versatile electron microscopy label. *Nat. Struct. Mol. Biol.* **17**, 775–778.

Gavin, A. C., Bosche, M., Krause, R., Grandi, P., Marzioch, M., Bauer, A., et al. (2002). Functional organization of the yeast proteome by systematic analysis of protein complexes. *Nature* **415**, 141–147.

Golas, M. M., Sander, B., Will, C. L., Luhrmann, R., Stark, H. (2003). Molecular architecture of the multiprotein splicing factor SF3b. *Science* **300**, 980–984.

Hainfeld, J. F. (1996). Labeling with nanogold and undecagold: techniques and results. *Scanning Microsc. Suppl.* **10**, 309–322.

Hainfeld, J. F., Safer, D., Wall, J. S., Simon, M., Lin, B., Powell, R. D. (1994). Methylamine vanadate (NanoVan) negative stain. In: Proc. 52nd Ann. Mtg. Micros. Soc. Amer., Bailey, G. W., Garratt-Reed, A. J., (Eds.), pp.132–133. San Francisco Press, San Francisco.

Han, B. G., Dong, M., Liu, H., Camp, L., Geller, J., Singer, M., et al. (2009). Survey of large protein complexes in D. vulgaris reveals great structural diversity. *Proc. Natl. Acad. Sci. USA* **106**, 16580–16585.

John, C. M., Hite, R. K., Weirich, C. S., Fitzgerald, D. J., Jawhari, H., Faty, M., et al. (2007). The Caenorhabditis elegans septin complex is nonpolar. *EMBO J.* **26**, 3296–3307.

Kastner, B., Fischer, N., Golas, M. M., Sander, B., Dube, P., Boehringer, D., et al. (2008). GraFix: sample preparation for single-particle electron cryomicroscopy. *Nat. Methods* **5**, 53–55.

Kato, T., Yoshida, H., Miyata, T., Maki, Y., Wada, A., Namba, K. (2010). Structure of the 100 S ribosome in the hibernation stage revealed by electron cryomicroscopy. *Structure* **18**, 719–724.

Kelly, D. F., Abeyrathne, P. D., Dukovski, D., Walz, T. (2008). The Affinity Grid: a pre-fabricated EM grid for monolayer purification. *J. Mol. Biol.* **382**, 423–433.

Kronenberg, S., Kleinschmidt, J. A., Bottcher, B. (2001). Electron cryo-microscopy and image reconstruction of adeno-associated virus type 2 empty capsids. *EMBO Rep.* **2**, 997–1002.

Kuhlbrandt, W. (1982). Discrimination of protein and nucleic acids by electron microscopy using contrast variation. *Ultramicroscopy* **7**, 221–232.

Kuhner, S., van Noort, V., Betts, M. J., Leo-Macias, A., Batisse, C., Rode, M., et al. (2009). Proteome organization in a genome-reduced bacterium. *Science* **326**, 1235–1240.

Lake, J. A., Pendergast, M., Kahan, L., Nomura, M. (1974). Ribosome structure: three-dimensional distribution of proteins S14 and S4. *J. Supramol. Struct.* **2**, 189–195.

Leschziner, A. E., Nogales, E. (2006). The orthogonal tilt reconstruction method: an approach to generating single-class volumes with no missing cone for ab initio reconstruction of asymmetric particles. *J. Struct. Biol.* **153**, 284–299.

Mastronarde, D. N. (2005). Automated electron microscope tomography using robust prediction of specimen movements. *J. Struct. Biol.* **152**, 36–51.

Mercogliano, C. P., DeRosier, D. J. (2007). Concatenated metallothionein as a clonable gold label for electron microscopy. *J. Struct. Biol.* **160**, 70–82.

Milligan, R. A., Safer, D. (1990). Molecular structure of F-actin and location of surface binding sites. *Nature* **348**, 217–221.

Muench, S. P., Huss, M., Song, C. F., Phillips, C., Wieczorek, H., Trinick, J., et al. (2009). Cryo-electron microscopy of the vacuolar ATPase motor reveals its mechanical and regulatory complexity. *J. Mol. Biol.* **386**, 989–999.

Munro, J. B., Sanbonmatsu, K. Y., Spahn, C. M., Blanchard, S. C. (2009). Navigating the ribosome's metastable energy landscape. *Trends Biochem. Sci.* **34**, 390–400.

Olson, N. H., Kolatkar, P. R., Oliveria, M. A., Cheng, R. H., Greve, J. M., McClelland, A., et al. (1993). Structure of human rhinovirus complexed with its receptor molecule. *Proc. Natl. Acad. Sci.* **90**, 507–511.

Pettersen, E. F., Goddard, T. D., Huang, C. C., Couch, G. S., Greenblatt, D. M., Meng, E. C., et al. (2004). UCSF Chimera—a visualization system for exploratory research and analysis. *J. Comput. Chem.* **25**, 1605–1612.

Puig, O., Caspary, F., Rigaut, G., Rutz, B., Bouveret, E., Bragado-Nilsson, E., et al. (2001). The tandem affinity purification (TAP) method: a general procedure of protein complex purification. *Methods* **24**, 218–229.

Radermacher, M. (1988). Three-dimensional reconstruction of single particles from random and nonrandom tilt series. *J. Electron Microsc. Tech.* **9**, 359–394.

Roberts, A. J., Numata, N., Walker, M. L., Kato, Y. S., Malkova, B., Kon, T., et al. (2009). AAA+ Ring and linker swing mechanism in the dynein motor. *Cell* **136**, 485–495.

Sagermann, M., Stevens, T. H., Matthews, B. W. (2001). Crystal structure of the regulatory subunit H of the V-type ATPase of Saccharomyces cerevisiae. *Proc. Natl. Acad. Sci. USA* **98**, 7134–7139.

Sander, B., Golas, M. M., Luhrmann, R., Stark, H. (2010). An approach for de novo structure determination of dynamic molecular assemblies by electron cryomicroscopy. *Structure* **18**, 667–676.

Sander, B., Golas, M. M., Stark, H. (2005). Advantages of CCD detectors for de novo three-dimensional structure determination in single-particle electron microscopy. *J. Struct. Biol.* **151**, 92–105.

Scheres, S. H., Nunez-Ramirez, R., Gomez-Llorente, Y., San Martin, C., Eggermont, P. P., Carazo, J. M. (2007). Modeling experimental image formation for likelihood-based classification of electron microscopy data. *Structure* **15**, 1167–1177.

Stark, H., Dube, P., Luhrmann, R., Kastner, B. (2001). Arrangement of RNA and proteins in the spliceosomal U1 small nuclear ribonucleoprotein particle. *Nature* **409**, 539–542.

Stewart, P. L., Fuller, S. D., Burnett, R. M. (1993). Difference imaging of adenovirus; bridging the resolution gap between X-ray crystallography and electron microscopy. *EMBO J.* **12**, 2589–2599.

Stroupe, M. E., Xu, C., Goode, B. L., Grigorieff, N. (2009). Actin filament labels for localizing protein components in large complexes viewed by electron microscopy. *RNA* **15**, 244–248.

Suhre, K., Navaza, J., Sanejouand, Y. H. (2006). NORMA: a tool for flexible fitting of high-resolution protein structures into low-resolution electron-microscopy-derived density maps. *Acta Crystallogr. D Biol. Crystallogr.* **62**, 1098–1100.

Suloway, C., Pulokas, J., Fellmann, D., Cheng, A., Guerra, F., Quispe, J., et al. (2005). Automated molecular microscopy: the new Leginon system. *J. Struct. Biol.* **151**, 41–60.

Tama, F., Miyashita, O., Brooks, C. L. III. (2004). Flexible multi-scale fitting of atomic structures into low-resolution electron density maps with elastic network normal mode analysis. *J. Mol. Biol.* **337**, 985–999.

Trus, B. L., Roden, R. B., Greenstone, H. L., Vrhel, M., Schiller, J. T., Booy, F. P. (1997). Novel structural features of bovine papillomavirus capsid revealed by a three-dimensional reconstruction to 9 A resolution. *Nat. Struct. Biol.* **4**, 413–420.

Ulbrich, C., Diepholz, M., Bassler, J., Kressler, D., Pertschy, B., Galani, K., et al. (2009). Mechanochemical removal of ribosome biogenesis factors from nascent 60 S ribosomal subunits. *Cell* **138**, 911–922.

Unwin, P. N. T. (1974). Electron-microscopy of stacked disk aggregate of tobacco mosaic-virus protein. 2. Influence of electron-irradiation on stain distribution. *J. Mol. Biol.* **87**, 657–664.

van Heel, M. (1984). Multivariate statistical classification of noisy images (randomly oriented biological macromolecules). *Ultramicroscopy* **13**, 165–183.

van Heel, M. (1987). Angular reconstruction: a posteriori assignment of projection directions from 3D reconstruction. *Ultramicroscopy* **21**, 111–124.

van Heel, M., Frank, J. (1981). Use of multivariate statistics in analysing the images of biological macromolecules. *Ultramicroscopy* **6**, 187–194.

Voss, N. R., Lyumkis, D., Cheng, A., Lau, P. W., Mulder, A., Lander, G. C., et al. (2010). A toolbox for ab initio 3-D reconstructions in single-particle electron microscopy. *J. Struct. Biol.* **169**, 389–398.

Wade, R. H. (1992). A brief look at imaging and contrast transfer. *Ultramicroscopy* **46**, 145–156.

Wang, G. J., Porta, C., Chen, Z. G., Baker, T. S., Johnson, J. E. (1992). Identification of a Fab interaction footprint site on an icosahedral virus by cryoelectron microscopy and X-ray crystallography. *Nature* **355**, 275–278.

Wilkens, S., Capaldi, R. A. (1992). Monomaleimidogold labeling of the gamma subunit of the Escherichia coli F1 ATPase examined by cryoelectron microscopy. *Arch. Biochem. Biophys.* **299**, 105–109.

Wolf, E., Kastner, B., Deckert, J., Merz, C., Stark, H., Luhrmann, R. (2009). Exon, intron and splice site locations in the spliceosomal B complex. *EMBO J.* **28**, 2283–2292.

Wriggers, W., Birmanns, S. (2001). Using situs for flexible and rigid-body fitting of multiresolution single-molecule data. *J. Struct. Biol.* **133**, 193–202.

Yoshioka, C., Pulokas, J., Fellmann, D., Potter, C. S., Milligan, R. A., Carragher, B. (2007). Automation of random conical tilt and orthogonal tilt data collection using feature-based correlation. *J. Struct. Biol.* **159**, 335–346.

Yu, X., Jin, L., Zhou, Z. H. (2008). 3.88 A structure of cytoplasmic polyhedrosis virus by cryo-electron microscopy. *Nature* **453**, 415–419.

Zhang, X., Settembre, E., Xu, C., Dormitzer, P. R., Bellamy, R., Harrison, S. C., et al. (2008). Near-atomic resolution using electron cryomicroscopy and single-particle reconstruction. *Proc. Natl. Acad. Sci. USA* **105**, 1867–1872.

Zhang, Z., Zheng, Y., Mazon, H., Milgrom, E., Kitagawa, N., Kish-Trier, E., et al. (2008). Structure of the yeast vacuolar ATPase. *J. Biol. Chem.* **283**, 35983–35995.

Zlotnick, A., Cheng, N., Stahl, S. J., Conway, J. F., Steven, A. C., Wingfield, P. T. (1997). Localization of the C terminus of the assembly domain of hepatitis B virus capsid protein: implications for morphogenesis and organization of encapsidated RNA. *Proc. Natl. Acad. Sci. USA* **94**, 9556–9561.

VISUALIZING MOLECULAR MACHINES IN ACTION: SINGLE-PARTICLE ANALYSIS WITH STRUCTURAL VARIABILITY

By SJORS H. W. SCHERES

MRC Laboratory of Molecular Biology, Cambridge, United Kingdom

Abstract

Many of the electron microscopy (EM) samples that are analyzed by single-particle reconstruction are flexible macromolecular assemblies that adopt multiple structural states in their functioning. Consequently, EM samples often contain a mixture of different structural states. This structural variability has long been regarded as a severe hindrance for single-particle analysis because the combination of projections from different structures into a single reconstruction may cause severe artifacts. This chapter reviews recent developments in image processing that may turn structural variability from an obstacle into an advantage. Modern algorithms now allow classifying projection images according to their underlying three-dimensional (3D) structures, so that multiple reconstructions may be obtained from a single data set. This places 3D-EM in a unique position to study the intricate dynamics of functioning molecular assemblies.

I. Introduction

Many vital processes in the cell are catalyzed by multicomponent macromolecular assemblies. These large complexes have also been called *molecular machines* because, just like machines invented by humans, they

ADVANCES IN PROTEIN CHEMISTRY AND STRUCTURAL BIOLOGY, Vol. 81
DOI: 10.1016/S1876-1623(10)81004-1

employ highly coordinated movements of separate parts to fulfill their tasks (Alberts, 1998). The understanding of how molecular machines work is a strategic goal in modern structural biology, but studying them is often challenging.

Proteomics studies have yielded a wealth of information on which individual proteins are present in these assemblies, and X-ray crystallography is delivering high-resolution structures for many of them. However, relatively little is known about how the different proteins interact to form functional assemblies. Fragile intermolecular interactions make it typically difficult to purify intact molecular machines, while different functional states may be hard to separate biochemically. Consequently, purified samples of molecular machines often suffer from various extents of nonstoichiometric complex formation and/or conformational variability. The occurrence of multiple different structures, also called *structural heterogeneity*, poses problems for many tools in structural biology. Structural heterogeneity often interferes with crystallization and tends to reduce the effectiveness of biophysical techniques that study complexes in bulk solution.

This chapter describes recent advances in image processing that have placed three-dimensional electron microscopy (3D-EM) in a unique position to study structurally heterogeneous samples of molecular machines. Modern electron microscopes allow the imaging of individual copies of these assemblies. Therefore, 3D-EM poses less stringent requirements on sample purity than many alternative techniques, provided that images of distinct 3D structures can be separated. Moreover, the complexes may be visualized in a thin layer of ice where they are free to adopt any of their functional states. In principle, one may therefore obtain structural information about a range of "snapshots" along the functional cycle of these machines from a single sample. Combining these snapshots into a movie of a functioning machine is then likely to teach us much more about its mechanisms than a single structure.

Still, the analyzing of structurally heterogeneous samples has only recently become possible. Not even a decade ago, structural heterogeneity was considered a strong limitation for the applicability of 3D-EM. If left untreated, structural heterogeneity in the images has detrimental effects on the structure determination process. At the very best, generalized isotropic variability may lead to an overall loss of resolution in an otherwise correct structural model, very much like the atomic motions that are modeled by temperature factors in X-ray crystallography would do. This kind of structural variability is likely to occur always to some extent and merely forms an additional resolution-limiting factor along with experimental noise, optical aberrations, detector imperfections, etc. Structural variability that is localized in a specific part of the complex is often more harmful. Flexible parts of a larger assembly may turn up as noninformative fuzzy densities, or their density may disappear altogether. Likewise, nonstoichiometric ligand binding may lead to a loss of density, possibly up to the point where the ligand can no longer be detected. Moreover, if this ligand binding induces structural changes in the rest of the assembly, these changes are likely to remain unobserved if images of ligand-bound complexes are not separated from unliganded assemblies. Very large conformational changes or large extents of compositional variability are potentially the most harmful form of structural heterogeneity in 3D-EM. If images from assemblies with major structural rearrangements, or even from completely different assemblies, are combined into a single 3D structure, the result may be completely artifactual. Unfortunately, many samples of molecular machines do suffer from relatively large extents of structural variability, which if left untreated, often prohibit their characterization by 3D-EM.

Because of its detrimental effects on the reconstruction process and because of a lack of suitable image-processing tools, one has conventionally aimed at avoiding structural heterogeneity in 3D-EM samples. The cryogenic temperatures that are used to freeze the samples reduce the intrinsic thermal vibrations of many flexible complexes. In addition, much like one might do for 3D crystallization experiments, one typically attempts to block dynamic molecular machines in a single structural state through the use of inhibitors, stabilizing mutations, modified substrates, etc. A further reduction of molecular flexibility or compositional variability may be obtained through chemical fixation. For example, glutaraldehyde or formaldehyde may be used to form stabilizing chemical bonds between flexible or unstable parts of the complexes (Wong and Wong, 1992). One typically assumes that this chemical cross-linking does not affect the structural interpretation at medium–low resolution (say below 10–15 Å), although in principle, one should be aware that a chemically altered structure is being analyzed. Recently, the introduction of a method that combines chemical fixation with the sedimentation of complexes in a density gradient has led to a marked increase of the use of chemical fixation in 3D-EM (Kastner et al., 2008).

Lately, the introduction of a range of image-processing tools for structurally heterogeneous samples has significantly relaxed the requirements of 3D-EM sample homogeneity. This chapter reviews these new techniques and illustrates their potential for the structural characterization of molecular machines. The methods described all fall under the so-called *single-particle* analysis, where projection images of many copies of structurally identical complexes with no or limited symmetry are combined into a 3D-reconstruction. Methods that make explicit use of higher order symmetry as in 2D crystals, helical, or icosahedral assemblies fall outside the scope of this chapter. Aimed at a general public of molecular biologists, this chapter explains the hurdles of single-particle analysis in the presence of structural heterogeneity and describes the general principles behind the methods that may overcome them. In addition, it discusses some as yet unrealized potential to convey part of the excitement that is felt among the 3D-EM community about studying molecular machines in their multiple functional states.

II. Two-Dimensional Image Analysis

Because biological macromolecules are easily damaged by electrons, the electron dose in the microscope needs to be rigorously limited. Consequently, EM images are typically very noisy and cannot be interpreted individually. The elevated noise levels may be reduced by averaging over many images. The resulting two-dimensional average images may then provide useful insights into the quality of the biological sample and the imaging conditions. As will be explained in more detail below, 2D averaging approaches on their own are of limited use in the analysis of structural heterogeneity. Still, many 3D classification approaches employ 2D averaging as a preprocessing step, and as such, 2D image analysis plays an important role in the analysis of structurally heterogeneous projection data.

It only makes sense to average over images with identical 2D signals. This means that even for a structurally unique 3D object, one can only average over projections of that object in the same 3D orientation (or "view"), as different views will give rise to different 2D signals. Even if a data set contains projections of a structurally homogenous sample in a single view, 2D averaging still requires image alignment to deal with random in-plane rotations and uncertainty in the image centers. This in-plane alignment is complicated by the abundant noise. Particularly for cryo-EM images, the high noise levels typically prohibit the alignment of a single experimental image to a second one. Better orientations may be obtained by aligning each noisy image to a noise-free reference image. But this represents a "chicken-and-egg" problem. The best possible reference image would be the correct underlying signal, but in order to obtain an estimate of this signal, the images first need to be aligned. To work around the dilemma, image-alignment approaches are typically iterative processes. An initial reference image of suboptimal quality may be used to align all images, the resulting alignment produces an improved average, this average is then used again as a reference for the next iteration, and so on (Frank, 2006).

Only very few (if any) data sets contain projections of only a single view of a structurally homogeneous molecule. In practice, most complexes adopt multiple orientations on the experimental support, and on top of that, many complexes display structural heterogeneity. Both effects give rise to distinct 2D projections that need to be separated. Therefore, apart from image alignment, 2D image analysis usually also involves a classification task to separate distinct 2D signals.

Also, image classification suffers from the high levels of noise. Again, two experimental images are typically too noisy to determine whether they belong to the same class or not. Fortunately, the noise levels may be greatly reduced by principal component analysis (PCA, also known as multivariate statistical analysis or MSA; van Heel and Frank, 1981; van Heel, 1984). These methods analyze the variance among all images in a data set and allow separation of the most significant structural differences (the so-called principal components) from differences caused by noise. The much less noisy principal components can be compared on an image-to-image basis in order to decide whether two images belong to the same class or not. Then, standard classification techniques like hierarchical ascendant or *K*-means clustering may be employed to separate the structurally distinct projections into different classes (see Frank, 2006 for more details).

However, another chicken-and-egg problem arises in image classification. If the images in a data set are not aligned, the principal components may reflect orientation rather than structural differences among the images. Therefore, one typically aligns images prior to their classification. But, as explained above, image alignment requires good reference images.

Ideally, one would need multiple references to accurately describe all 2D structures in the data, but these could only be obtained if the alignment and classification problems were already solved. In this case, iterative procedures also provide a workaround. All images may initially be aligned against a single reference and subsequent classification may yield multiple image subsets. Separate alignment of these subsets may then yield improved alignments that may be used to improve classification. A different option is to align the images through the classification procedure itself. In the so-called *alignment-through-classification* approach (Dube et al., 1993), PCA is used to group the unaligned images according to similarity in their orientations (as well as their underlying 2D structures).

Alternatively, image alignment and classification may be tackled in a single optimization process. Similar to image alignment, one may classify noisy images by comparison with multiple, relatively noise-free reference images. If one uses multiple references inside the alignment procedure, one can simultaneously align and classify the noisy images (van Heel and Stoffler-Meilicke, 1985). A particularly powerful approach to this *multireference alignment* scheme was recently introduced in the form of a maximum likelihood (ML) algorithm (Scheres et al., 2005b). The ML formulation provides a powerful statistical framework for these ''chicken-and-egg'' type of problems and has been shown to be particularly robust to high levels of noise (Sigworth, 1998). Rather than assigning optimal orientations and classes to each experimental image, the ML approach calculates probabilities for all possible assignments and then calculates class averages using all probability-weighted assignments (see Sigworth et al., 2010 for a recent review on ML approaches in the field). It was shown that 2D ML multi-reference alignments may be started from average images of random subsets in random orientations, making the procedure completely unsupervised.

In principle, if a projection data set of a structurally heterogeneous particle is divided into a sufficiently high number of 2D classes, each 2D class would contain projections of only a single 3D structure. Still, this is typically not enough to solve the problem of 3D reconstruction for each of these structures. The differences among the 2D class averages do not originate only from the different 3D structures, but also from the different views that each structure adopts on the experimental support. The main problem with 2D image analysis is the difficulty to distinguish these two types of differences: how are the 2D class averages grouped together based on their underlying 3D structure? A large variability in size, possibly related to aggregation or sample impurity, could perhaps be dealt with in 2D. But more often, the structural differences among the class averages are rather subtle and it is very difficult to decide whether these differences originate from different views or from structural heterogeneity (see Fig. 1). Therefore, based on 2D analysis alone, it is typically very difficult to draw conclusions about the structural homogeneity of a sample.

III. Three-Dimensional Image Analysis

If multiple views of a unique 3D object are available, one may reconstruct its 3D structure using computerized techniques. As in 2D analysis, many images of a unique object need to be combined in order to reduce the noise, and the relative orientations of all images need to be known. When the data contain projections of multiple 3D structures, the problem may again be expressed in terms of a combined (3D) alignment and classification task. As mentioned above, the main difficulty in 3D classification is how to tell different views of a unique structure from projections of different structures. In addition, 3D reconstruction itself is mathematically much more complex than 2D averaging. Even when many projections of a structurally unique 3D object are available and their relative orientations are known, 3D reconstruction is a complicated task. It is therefore not surprising that the same types of chicken-and-egg problems described in the previous section play even more important roles in 3D.

Also, the process of alignment is more complicated in 3D than in 2D. Even when all images in a data set originate from a structurally unique object, the 2D projections cannot be compared directly to each other because different views give rise to structurally different images. Consequently, 3D alignment depends even more than its 2D counterpart on the use of reference images. If an initial 3D reference structure is available, one may calculate a library of reference projections of this object in all possible views. (Methods for obtaining an initial model *de novo* are available; see Frank, 2006 for more details.) Comparison with each of the reference library projections may then serve to assign 3D orientations to all experimental images. Obviously, this procedure suffers from a similar type of chicken-and-egg problem as 2D alignment. The best reference projections would originate from the correct 3D object, but this structure cannot be calculated as long as the images are not aligned.

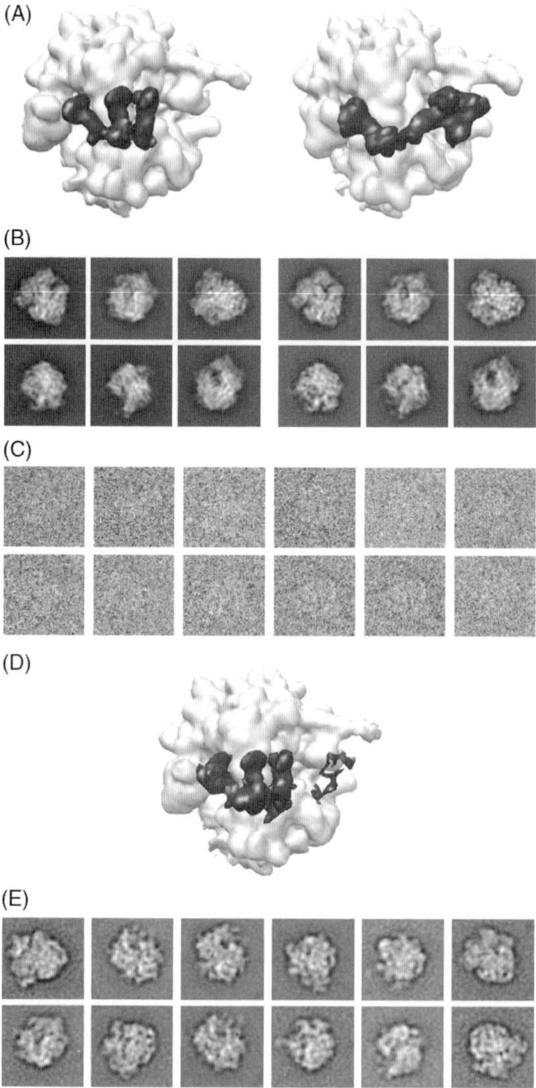

Fig. 1. *An example of structural heterogeneity.* (A) Two 70S *E. coli* ribosome structures. The left-hand side structure contains three tRNA molecules and the right-hand structure contains one tRNA molecule and elongation factor G (EF-G). Ribosomal density is shown in gray, tRNA and EF-G densities are superimposed in black. (B) Computer-generated projections of the two ribosome structures. (C) Experimental cryo-EM projection images from a sample that contains the two structures shown in (A). (D) A single 3D reconstruction made from a structurally heterogeneous set of projection images originating from the two structures shown in (A). Density for the EF-G ligand is very weak and unconnected due to a low percentage of the images originating from the structure with EF-G. (E) Two-dimensional class averages from the structurally heterogeneous data set used to calculate the structure shown in (D). It is very hard to tell to which 3D structure each 2D class average belongs (cf. B).

Again, iterative procedures provide a workaround. An initial, suboptimal 3D reference will likely result in suboptimal alignments, but subsequent reconstruction may then yield an improved model for a next iteration. However, because of the mathematical complexity of the reconstruction problem and the elevated noise levels in the data, this refinement process suffers to a large extent from *bias* toward the initial model. This bias makes it hard to remove incorrect features in the reference, which makes the outcome of the process highly dependent on the choice of the initial model.

The problem is even more complex if the sample is structurally heterogeneous. In that case, separate reconstructions should be performed for each of the different 3D structures in the data. As in 2D analysis, class assignments could again be made based on comparison with multiple references, but this once again represents a complicated chicken-and-egg problem. Only 3D structures that represent the structural variability in the data would yield useful reference projections, and these reconstructions cannot be calculated as long as the 3D alignment and classification problems have not been solved.

The remainder of this section gives an overview of the image-processing tools that deal with the problem of 3D reconstruction from structurally heterogeneous projection data. Before describing methods to separate structural heterogeneity, a tool to detect and localize structural variability in 3D reconstructions is discussed. Then, the pros and cons of existing classification methods are discussed according to a division into three categories. The first category concerns *supervised classification,* which has been the oldest approach to structural heterogeneity in the field. Supervised classification relies on prior information about the structural heterogeneity to generate two or more different 3D reference structures. Projections of these references then serve to assign both orientations and classes to all experimental images. The second category comprises *unsupervised classification* approaches, which have only recently become available. Unsupervised methods do not depend on prior knowledge about the structural variability in the sample, although most of them do rely on a single 3D reference structure for initial alignment. This structure should, to some extent, represent the range of structures in the data and will be called the *consensus structure.* Many unsupervised approaches depend on conventional methods to obtain class averages and relate these to 3D conformations in various ways, while other unsupervised approaches solve the problem intrinsically in 3D. The third category will be referred to as *tomographic classification.* This approach exploits pairs or series of images that are recorded at different tilt angles in the microscope to generate multiple 3D reconstructions that represent the structural variability in the particles.

A. *Detecting Structural Heterogeneity*

The detection of structural heterogeneity is not trivial. Observations that a 3D reconstruction is of lower quality than expected are often subjective and, as explained above, 2D class averages are a poor indicator. A potentially quantitative tool for the detection and localization of structural heterogeneity is the 3D variance map. If one would be able to calculate a large number of independent 3D reconstructions from the data, one could calculate the corresponding variance among all these reconstructions. Structurally variable regions of the molecules would then show up as regions with high intensity in the 3D variance map, while analysis of the covariance among all reconstructions could serve to obtain additional information about correlated structural changes.

Perhaps the most straightforward way to calculate a 3D variance map would be to divide the data into many subsets and to refine each subset separately using a consensus structure as an initial model. But in this scenario, the original data set would need to be very large because each reconstruction requires a minimum number of images to reduce the noise. Penczek et al. (2006b) proposed a more efficient method to calculate 3D variance based on *bootstrapping.* In this method, one generates a large number of subsets by omitting randomly selected images from the original data set while using others more than once (see Fig. 2). In that way, many subsets of sufficient size may be obtained from commonly sized data sets. It was shown for both model data and experimental data on ribosome complexes that 3D variance maps of reconstructions obtained by the bootstrap method yield useful information about the presence and location of structural heterogeneity (Penczek et al., 2006a). Recent improvements to this technique include a fast reconstruction algorithm that takes the transfer function of the microscope into account, and a validation method that determines the minimum number of reconstructions needed (Zhang et al., 2008).

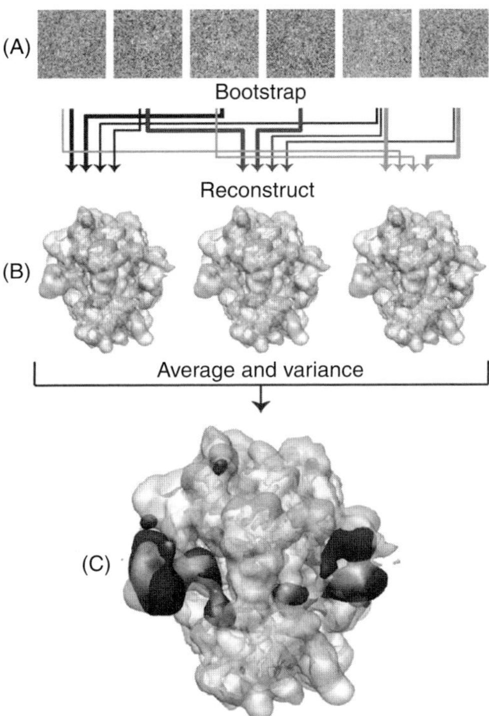

FIG. 2. *The bootstrap method to calculate 3D variance.* In the bootstrap method, the possibly structurally heterogeneous data set (A) is resampled into many (typically hundreds) of randomly different data sets. This random sampling occurs with replacement, which means that an image may be used more than once (indicated with thicker lines), while others are not used in a given resampled set. Using previously assigned orientations (based on alignment against a consensus model), for each of the resampled sets, a 3D reconstruction is calculated (B). The resulting maps are then used to calculate an average and variance map (C; transparent gray and black, respectively). Structurally variable regions show up as intense peaks in the variance map.

Currently, the 3D variance map is the only available tool to detect structural heterogeneity in 3D reconstructions. A potential drawback of the approach is that the angular assignments that are required for the multiple 3D reconstructions are based on alignment of the entire data set against a single consensus structure. If the structural heterogeneity is large, this consensus structure may likely be a poor model for some of the conformations in the data. In that case, incorrect angular assignments may result in systematic errors in the 3D variance map. To date, it is not clear to what extent these errors affect the applicability of the method in the general case.

B. Supervised Classification

If one knows where to look, finding something is not so difficult. Supervised classification uses prior knowledge about the structural variability to generate two or more 3D reference structures. Comparison with projections of each of these 3D references in all possible views serves to both align and classify the experimental images. Optionally, this method may be applied in an iterative manner where images assigned to each reference are used to reconstruct new references for a next iteration. In its iterative form, this procedure is also called *competitive projection matching.*

In the past, supervised classification has been applied on several occasions to data sets of the 70S *Escherichia coli* ribosome, for example (Valle et al., 2002; Gao et al., 2004; Agirrezabala et al., 2008). During the elongation cycle of protein synthesis, the smaller 30S subunit of this complex rotates with respect to the larger 50S subunit, the so-called ratchet movement. In addition, depending on its ratcheted state, the ribosome may bind various factors and tRNA molecules in distinct conformations. Therefore, reconstructions with blurred density for the 30S subunit together with low density for some of the factors may indicate possible mixtures of unratcheted and ratcheted ribosomes. In such cases, useful reference structures may be obtained from previously solved structures of ribosomes in unratcheted and ratcheted states. Often, in an attempt to avoid model bias, all factor density are removed from these maps. Similarly, Heymann et al. (2003) used linear combinations of procapsid and mature virus reconstructions to generate multiple 3D references in a supervised classification approach to study HSV-1 and HK97 virus maturation processes.

Obviously, the applicability of the supervised approach is limited because prior knowledge about the variability in the data is not available in the general case. An early workaround to this problem aimed to ''predict'' the structural variability from a single, low-resolution consensus model based on a highly simplified physical model of protein structures. In this approach, low-resolution reconstructions are modeled as ''beads connected by springs'', and normal mode analysis from molecular dynamics is employed to predict the most likely conformational changes (Tama et al., 2002). Based on a 19 Å consensus structure, Brink et al. (2004) predicted possible conformational changes for fatty acid synthetase and refined the predicted models in a supervised classification approach. The refined models were similar to the predicted ones. To rule out model bias in this refinement, they also refined each class separately using the consensus model as initial reference. Each of these refinements resulted in a structure that was similar to the corresponding one from the supervised approach. Although normal mode analysis of this type has been applied to predict molecular motion on some occasions (e.g., also see Tama and Brooks, 2002), it is not clear how the highly simplified physical model may describe molecular flexibility in a general way.

The most important drawback of supervised classification is its strong dependence on initial references, which makes it particularly vulnerable to the dangers of model bias. If the assumptions or predictions about the structural variability in the data are incorrect, this method may yield completely meaningless solutions, which may be hard to discern from the correct one. This is the main reason why nowadays unsupervised methods are generally preferred over the supervised ones.

C. *Unsupervised Classification*

The recent introduction of various unsupervised classification approaches is the main cause for optimism in the field about the possibilities to study flexible molecular machines by 3D-EM. Because unsupervised classification methods do not depend on prior knowledge about the structural variability in the data, they are applicable to many more cases than the supervised approach.

Various methods have been proposed that rely on conventional 2D analysis techniques to classify the images in a preprocessing step and then relate the 2D class averages to distinct 3D conformations in different ways.

Orlova and colleagues proposed the use of PCA to separate overall size differences among particles (White et al., 2004). In that case, some of the principal components of the variance among all images (also called *eigenimages*) were shown to display concentric rings that represent the variation in size of projections in all possible orientations. Classification based on these components of the variance was shown to successfully separate projections of small heat shock protein Hsp26, with size differences as small as 5–10% (White et al., 2006). A similar analysis may also serve to identify those variations among the data that are caused by the presence or absence of a ligand, which typically show up as intense, localized peaks in the eigenimages (Elad et al., 2008). Classifications based on such eigenimages were used to separate complexes of groEL with or without ligand or the 70S *E. coli* ribosome with or without elongation factor G (EF-G) (Elad et al., 2007). A drawback of these approaches is that their outcome may depend strongly on the expertise of the person who selects and analyzes the eigenimages. Moreover, their applicability in cases without an overall size difference or the presence or absence of a ligand is unclear.

An alternative technique to separate ligand-bound from unbound complexes employs the 3D variance map to localize structural variability and then focuses classification to the corresponding areas in the 2D projections (Penczek et al., 2006a). A nonstoichiometrically bound ligand will show up as an intense peak in the 3D variance map. A 3D mask that encapsulates this peak may be projected in all directions to identify the position of the ligand in the 2D projections. Ligand-bound particles may then be separated from unbound particles because the former will give rise to higher density in that region. The efficiency of this so-called *focused classification* method was illustrated for a ribosome complex with or without EF-G. In cases of domain movement or generalized conformational changes, the applicability of this approach may be limited. Although the area of variability may be identified in the 3D variance map, relating the class averages to the different 3D structures may be difficult because different projection directions may give rise to either a gain or a loss of density in the corresponding 2D areas. Nevertheless, this technique has been successfully applied to a number of different systems, including yeast fatty acid synthase (Gipson et al., 2010) and RNA polymerase I (Kuhn et al., 2007) and II (Kostek et al., 2006).

Two more approaches that use 2D class averages exist, but they are not restricted to a specific type of structural variability. Both approaches use a 3D consensus structure to assign initial orientations and then apply standard 2D classification approaches to groups of experimental images with similar views. Firstly, Fu et al. (2007) exploit the observation that projections from a unique 3D object in a small angular neighborhood (i.e., with similar views) will give rise to similar images. Inside this neighborhood, it is then possible to distinguish distinct conformations by standard 2D classification procedures, and one may "track" the distinct 3D classes by progressively moving along the angular projection space. Obviously, this *cluster tracking* method relies heavily on a sufficient sampling of 2D class averages over the full angular coverage. This requirement is often hard to meet in practice, which may explain why no successful applications of this method have been reported yet.

Secondly, Hall et al. (2007) proposed a method that relates 2D class averages to 3D conformations in a more widely applicable manner. Their method makes use of the *central section theorem*, which states that Fourier transforms of projections of a 3D object correspond to central sections of the 3D Fourier transform of that object. Thereby, all projections from a unique 3D object share a line in Fourier space, the so-called *common line*, while class averages that originate from distinct 3D objects do not share a common line. Because the class averages are calculated for subsets of images with identical, known orientations with respect to the consensus model, the orientation of the common line for all pair-wise combinations of class averages in distinct orientations is known. Thereby, the class averages may be separated according to their underlying 3D structures based on maximizing the cross-correlation between common lines within each 3D class. This so-called *cross-correlation of common lines* (CCL) method was recently automated by expressing the optimization task as a graph-cut problem from standard graph theory (Shatsky et al., 2010). In this approach, a graph is constructed where all class averages are connected to one another by edges that are weighted according to the cross-correlation of their common line (see Fig. 3). The algorithm employed aims at cutting the graph into a user-defined number of subgraphs, such that the weights of edges within each subgraph is maximized (or the weights of edges between different subgraphs are minimized). The efficiency of the approach was illustrated for 70S *E. coli* ribosomes with or without EF-G, human translation initiation factor eIF3 with or without an internal ribosome entry site (IRES) ligand, and different open and closed conformations of human RNA polymerase II.

Previously, Herman and Kalinowski (2008) had proposed a different classification approach that was also expressed as a graph-cut problem and where the weights of the graph's edges were also calculated based on common line similarities. Interestingly, neither did these authors determine any relative orientations nor did they use any 2D averaging. Instead,

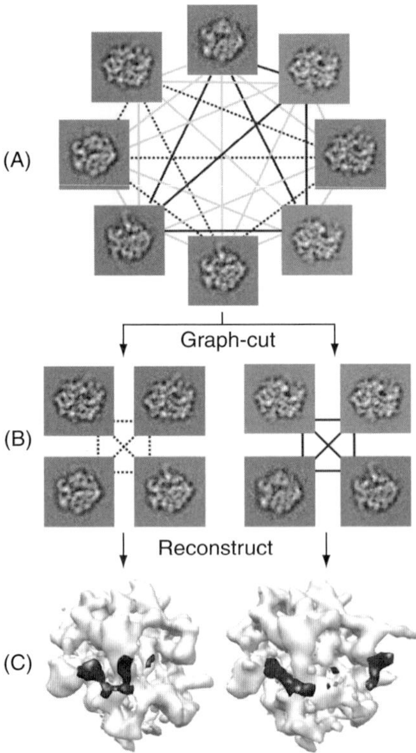

FIG. 3. *Automated cross-correlation of common lines.* In two preprocessing steps, all experimental images are aligned against a single, consensus model, and standard 2D classification tools are used to divide images with identical projection angles into different classes. Then, a weighted graph (A) is constructed where every 2D class average is connected with all other 2D class averages with different projection angles. The weights of the graph are calculated from the cross-correlation coefficients of the common lines between each pair of class averages. Pairs of class averages originating from the same structure are shown in black (using continuous or dotted lines for the two different structures). Pairs originating from different structures are shown in gray. The structural heterogeneity is separated by cutting the graph into two subgraphs (B) so that the intragraph weights are maximum or the intergraph weights minimum. For each subgraph, a 3D reconstruction is calculated using the originally assigned projection angles (C).

they performed a search for the best possible common line between all pairs of individual, experimental images. The resulting graph is much larger than in the CCL approach, but it was demonstrated that cutting this graph can still be performed very efficiently. While the information content of common lines is notoriously low, particularly for individual images, the simultaneous consideration of all possible image pairs was hypothesized to make this approach relatively robust to the high noise levels. For simulated noisy data on geometrical phantom structures, this approach was indeed shown to yield highly pure classes, but its efficiency was never demonstrated for experimental data.

On their own, many of the methods described above typically yield classification solutions of only moderate quality (e.g., see Fig. 3). Suboptimal classifications may be caused by a combination of factors related to the various processing strategies. Conventional 2D class averaging approaches may yield relatively impure classes, common lines may carry too little information to allow reliable class assignments, or the use of a single consensus model may result in suboptimal angular assignments in the presence of large conformational changes. Still, as long as the differences between the resulting 3D class reconstructions resemble the structural variability in the sample to a large enough extent, the classes may be improved significantly through subsequent competitive projection matching. For example, refinement of the maps shown in Fig. 3 leads to two ribosome structures that are nearly indistinguishable from the ones shown in Fig. 1A (results not shown).

The potential of competitive projection matching to improve initial classes of moderate quality was taken to an extreme by Scheres et al. (2007b). These authors showed that a ML formulation of multireference refinement (called *ML3D*) may converge to highly pure classes even when starting from random classifications. Similar to the ML2D multireference alignment approach discussed above, in the ML3D method, one calculates probabilities for all possible orientation and class assignments and one performs 3D reconstructions by considering all probability-weighted assignments. The only requirement of this approach is the availability of a very low-resolution (e.g., 80 Å) consensus model. This model is refined for a single iteration against random subsets of the data in order to generate a user-defined number of reference maps. These maps are then refined simultaneously against the entire data set for multiple iterations of competitive projection matching based on the ML3D algorithm (see Fig. 4). It was demonstrated that the radius of convergence of this statistical refinement procedure is sufficiently large to separate distinct conformations: 70S ribosomes in complex with EF-G were separated from

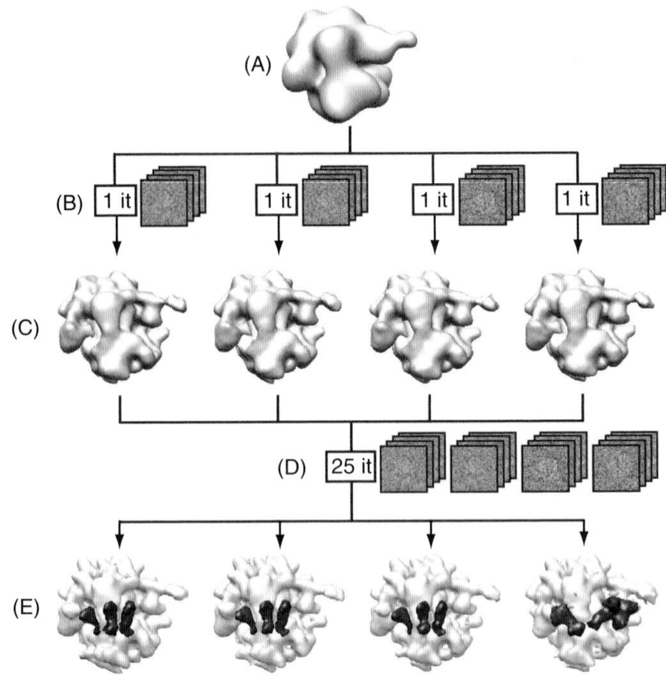

Fig. 4. *ML3D classification.* A low-pass filtered consensus model (A) is used as a single reference for one iteration of 3D maximum likelihood refinement against (in this case) four random subsets of the structurally heterogeneous data set (B). This yields four structures with random differences between them (C). These structures are used as initial models for a multireference ML refinement against the entire data set (D). After 25 iterations, the ML refinement converges to a solution where the different reference structures represent the structural variability in the data (E).

ribosomes without EF-G, and Simian Virus 40 large T antigen particles were classified according to an overall bending movement.

Subsequent improvements of the ML method comprised better statistical models for 3D-EM images (Scheres et al., 2007a, 2009b; Scheres and Carazo, 2009) and efforts to reduce its relatively high computational costs. Although the computational costs originally represented a serious drawback of the ML approach, currently available modified algorithms are one or two orders of magnitude faster than the original approach (Scheres et al., 2005a; Tagare et al., 2008, 2010) and can be run in a few days on commonly available high-performance computing hardware. The ML3D approach enjoys the general applicability of competitive projection matching, but it does not require initial references that reflect the structural variability in the data. This makes the ML3D approach applicable to many systems, as is also illustrated by its application to a wide variety of structures, among others, chaperonin CCT (Cuellar et al., 2008), the 26S proteasome (Nickell et al., 2009), the human RISC-loading complex (Wang et al., 2009), and yeast DNA polymerase alpha (Klinge et al., 2009).

D. *Tomographic Classification*

The intricate entanglement of orientation and class assignments may be partially resolved through the use of a more elaborate data collection strategy. When images are collected in pairs (or series) at different tilt angles in the microscope, one knows that images within a pair belong to the same 3D structure. In addition, the relative tilt angles at which these images were recorded are also known. This prior information greatly facilitates 3D alignment and classification.

In the *random conical tilt* (RCT) technique, one records pairs of images at different tilt angles in the microscope (Radermacher et al., 1986; Radermacher, 1988). For each field of view, one takes one image at a relatively high tilt angle (often in the range of 40–55°) and one image without tilting. Classification and part of the alignment can then be tackled by applying standard 2D class averaging techniques on the untilted particles. The remaining part of the alignment problem is solved through geometrical considerations about the experiment. As a result, one may calculate a single 3D reconstruction for each of the 2D class averages from the untilted images. An example application that illustrates the potential of this technique for structurally heterogeneous samples was recently described by Radermacher (2009), who used this technique to obtain multiple 3D models for the highly dynamic mitochondrial Complex I from *Yarrowia lipolytica*.

Using the RCT technique, two class averages that correspond to different views of the same 3D object in the untilted images will give rise to two similar reconstructions in different orientations. Likewise, two class averages that originated from different 3D objects will give rise to structurally different 3D reconstructions. Therefore, the problem of separating structural heterogeneity may be reformulated as a 3D averaging task that requires both alignment and classification of 3D maps. Apart from its increased dimensionality, this problem is very similar to 2D averaging and is as such much easier to solve than the 3D reconstruction problem from structurally heterogeneous 2D projections.

However, because of the limited tilt angle in the microscope, the distribution of 3D orientations of images in a RCT reconstruction does not cover the full angular range. Consequently, RCT reconstructions contain empty regions in Fourier space, which cause significant artifacts to the real-space maps and complicate their alignment and classification. Alternatively, if the particle adopts a relatively large number of orientations on the experimental support, one may use the *orthogonal tilt reconstruction* (OTR) method (Leschziner and Nogales, 2006). This method yields reconstructions that are free from artifacts by recording tilt pairs at $-45°$ and $+45°$. Using this technique, Leschziner et al. (2007) obtained models for an open and closed conformation of the yeast chromatin remodeling complex RSC.

If tilt series rather than tilt pairs are recorded, one may calculate 3D reconstructions for entire fields of views in the microscope. These tomographic reconstructions may contain multiple copies of a molecular machine in different orientations or conformations. One may extract these complexes as individual 3D particles and reduce their noise levels through averaging. This process of *subtomogram averaging* falls outside the scope of this paper, but it is also mentioned here because of its overlap with the averaging of RCT reconstructions. Again due to limited tilt angles in the microscope, subtomograms also contain empty regions in Fourier space, which results in similar artifacts as in RCT reconstructions. Moreover, the tasks of 3D alignment and classification of both types of maps are identical. Consequently, algorithms for the averaging of RCT reconstructions may be easily adapted for subtomogram averaging, and vice versa.

Recently, three similar approaches that express the task of alignment and/or classification of RCT reconstructions in a statistical framework have been proposed. Firstly, Scheres et al. (2009a) proposed a ML multi-reference refinement scheme for the simultaneous alignment and classification of 3D maps with empty regions in Fourier space. Not only does the ML formulation provide a powerful way to deal with the alignment and classification, it also provides a statistical framework to deal with the empty regions in Fourier space. It was shown for subtomograms of groEL and groEL:groES complexes and for RCT reconstructions of human tumor suppressor protein p53 that the radius of convergence of this approach is large enough to start refinements from averages of random subsets of the 3D particles in random orientations. Thereby, this approach may align and classify subtomograms or RCT reconstructions without any prior knowledge about the structures or structural variability in the data.

Secondly, Sander et al. (2010) proposed a similar algorithm to align RCT reconstructions through a weighting scheme that was also inspired by ML principles. Again, an exceptionally large radius of convergence was reported, illustrating that model bias plays a much reduced role in the averaging of RCT reconstructions. Not only could RCT reconstructions be aligned against an initial random noise model, even initial models with the wrong hand or a completely different structure could be refined successfully. Subsequent to alignment, structural variability in the RCT reconstructions was classified using 3D PCA and standard hierarchical clustering techniques (see Fig. 5). The combination of the weighted alignment scheme with classification based on 3D PCA was shown to yield *de novo* reconstructions for different conformations of 70S *E. coli* ribosomes and human U4/U6.U5 tri-snRNP complexes (Sander et al., 2006).

Thirdly, Yu et al. (2010) proposed an improved version of 3D PCA for maps with empty regions in Fourier space. Also, these authors use an ML formulation, in this case, to find the principal components of the variance in the RCT reconstructions. In this method, the empty regions in Fourier space are treated in a very similar manner as in the approach by Scheres et al. (2009a). The efficiency of the method was illustrated for simulated data and for reconstructions of experimental images on *Saccharomyces cerevisiae* phosphofructokinase with simulated empty regions in Fourier space.

IV. Outlook

Traditionally, the ribosome has been a favorable sample for EM. Its large size (~2.5 MDa) and high RNA content (which scatters electrons more strongly than protein does) give rise to images with relatively high contrast and signal-to-noise ratios. Ribosomes were among the first samples to be studied by single-particle 3D-EM, and the problems encountered in these pioneering studies have been the driving force for many of the currently available image-processing tools. It is therefore not surprising that ribosome data sets have also played a central role in the development of algorithms to classify structural heterogeneity.

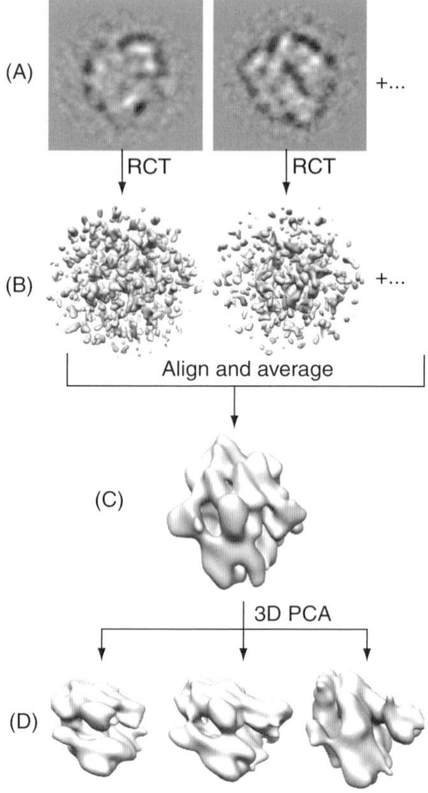

FIG. 5. *Weighted averaging of RCT reconstructions.* From a relatively large data set of tilt pairs, multiple 2D class averages (with ~40 images per class) are calculated for the untilted images (A). For each class, a (very noisy) RCT reconstruction is calculated (B). Reference-free alignment and averaging of all RCT reconstructions is performed using a weighting scheme that accounts for the different numbers of experimental images in each class. The resulting average shows a ribosome structure (C), and different conformations may then be separated by 3D PCA on the aligned RCT reconstructions (D).

The availability of many different ribosome structures has facilitated early supervised classification approaches, while more recently, the efficiency of many unsupervised approaches has been demonstrated using ribosome data (also see Figs. 2–5). This may partially be explained by the introduction of a standard ribosome data set for the testing of new 3D classification approaches (Baxter et al., 2009). However, the intrinsic flexibility of the ribosome often results in structurally heterogeneous samples, and multiple ribosome data sets with structural heterogeneity have now been described. Some of these data sets have driven the development of the classification approaches described above, while others were classified using existing tools (e.g., Cheng et al., 2010; Weis et al., 2010).

Nowadays, 3D-EM is routinely applied to a wide variety of macromolecular machines, and it is becoming increasingly clear that molecular flexibility plays a crucial role for many of these complexes. Structural heterogeneity in 3D-EM samples may well be the rule rather than the exception. It is therefore particularly promising that various 3D classification approaches discussed in this chapter have been applied to a range of different complexes. A small selection of reported classifications is illustrated in Fig. 6. The variety of methods employed and the different types of structural variability classified illustrate the potential of general applicability of the 3D-EM approach to structurally heterogeneous samples.

In the examples in Fig. 6, the data sets were separated into a relatively small number of classes (most typically, two or three), which represents a general trend in recently reported classifications. But if molecular machines really are flexible, they are likely to employ continuous domain movements that cannot be captured in only a few snapshots. In that case, in order to make a "molecular movie" as referred to in the introduction, many structures should be reconstructed from a single sample. As a certain amount of images need to be combined in each reconstruction in order to reduce the noise, a large number of structures would call for very large data sets. Cryo-EM single-particle reconstructions often contain several tens of thousands of images, so that data sets of several millions of particles will probably be necessary in order to reconstruct many structures. Recent efforts to automate 3D-EM data collection (Suloway et al., 2005; Stagg et al., 2006; Nickell et al., 2007) will play crucial roles in recording these large data sets.

Again, it is the ribosome that plays a pivotal role in moving the 3D-EM field to the next stage. Recently, Stark and colleagues collected a data set comprising 2 million 70S ribosome particles (Fischer et al., 2010). Whereas ribosome samples are typically prepared at 4 °C, these authors prepared their samples at 18 °C prior to freezing for cryo-EM. Using a hierarchical strategy combining both supervised and unsupervised classification techniques, the data set collected from this sample was then separated into 50 different classes. Combined into a movie, the corresponding structures

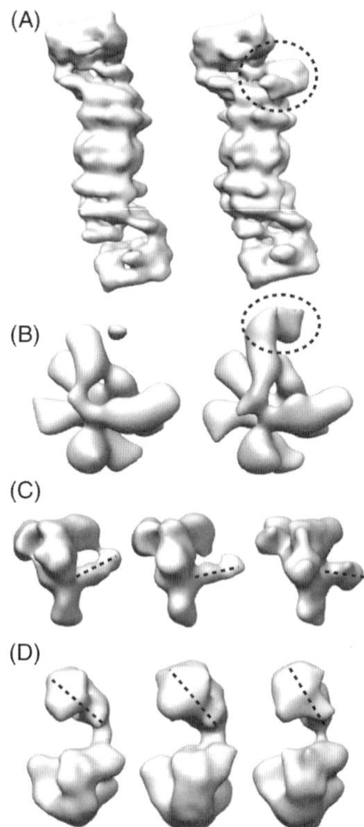

FIG. 6. *Example classifications of asymmetric molecules.* (A) Cryo-EM images of *Drosophila melanogaster* 26S proteasomes were classified using the ML3D algorithm (Scheres et al., 2007b). The two distinct classes differ in the presence or absence of an additional mass (indicated with an ellipse) near the entrance of the ATPase ring (Nickell et al., 2009). (B) Negative stain images of translation initiation factor eIF3 bound to hepatitis C virus internal ribosome entry site (IRES) were classified using the automated cross-correlation of common lines method. The two classes reveal unbound eIF3 molecules as well as eIF3 molecules in complex with IRES (indicated with an ellipse; Shatsky et al., 2010). (C) Cryo-EM tilt pair images of human U4/U6.U5 tri-snRNPs were classified using weighted RCT reconstruction averaging. The different classes reveal different conformations of the stalk (indicated with dashed lines; Sander et al., 2010). (D) Negative stain images of yeast DNA polymerase α were classified using the ML3D algorithm (Scheres et al., 2007b). The resulting three classes provide snapshots of a supposedly continuous degree of bending between the catalytic domain (below) and B subunit (top; Klinge et al., 2009).

reveal the passing of tRNA molecules through the ribosome, accompanied by the ratchet movement of the ribosome. These observations are in agreement with previous single-molecule FRET experiments (Cornish et al., 2008) and cryo-EM reconstructions (Agirrezabala et al., 2008; Julian et al., 2008) that indicated that the thermal energy in the ribosome system is large enough to induce such large conformational changes. Although the partially supervised character of the classification approach makes it difficult to predict its applicability for systems other than the ribosome, this experiment does illustrate the exciting potential of 3D-EM to study dynamic molecular machines.

Still, there is ample room for improvement of the methods involved. As illustrated in Section III.D, pairs or series of tilted images have great potential for *de novo* structure determination of highly dynamic molecular machines. However, this potential has not been fully exploited yet. Large data sets of tilt pairs are rare because of the experimental difficulties involved, but ongoing automation efforts (e.g., Yoshioka et al., 2007; Suloway et al., 2009; Voss et al., 2009) are likely to reduce these difficulties. Adaptation of existing refinement programs to handle large sets of tilt pairs or series may then greatly reduce the problems of model bias in the single-particle approach. Moreover, apart from adding valuable prior information about the relative orientations to the refinement, tilt pairs may also be used as a means to validate the structures obtained (Rosenthal and Henderson, 2003). The latter may play an important role in relieving the general concerns in the field about the lack of an equivalent of the crystallographic free R factor (Brunger, 1992).

In addition, it is likely that statistical approaches, such as the ML method, will continue to play increasingly important roles as statistical data model provide more robust ways of handling the high levels of experimental noise in 3D-EM images. Further improvements to these approaches are expected from the development of more accurate data models or through the incorporation of prior knowledge in *maximum a posteriori* approaches (also see Sigworth et al., 2010). But perhaps the biggest improvements in single-particle analysis may be expected from hardware rather than software developments. Direct electron detectors (Faruqi, 2009), possibly suitable for single-electron counting (McMullan et al., 2009), together with improved electron optics like aberration correctors (Rose, 2009) or phase plates (Nagayama and Danev, 2008), are expected to significantly improve cryo-EM image contrast and signal-to-noise ratios in the coming years. Also, the development of devices that allow time-resolved 3D-EM experiments by quickly freezing samples at precise time points after initiating a reaction are promising tools for the characterization of dynamic molecular machines (Berriman and Unwin, 1994; Moffat and Henderson, 1995; White et al., 2003; Shaikh et al., 2009).

It is difficult to predict what the impact of these combined developments on the field will be. One could envision a scenario where molecular movies may be obtained for a wide range of dynamic machines. Taken to the extreme, improved 3D-EM image classification approaches may even replace part of the biochemical sample purification process. Relatively impure cell extracts might be imaged directly in the electron microscope, provided that particles of interest may be separated from the rest of the sample by image classification. But even in a less optimistic view, continuing developments in dealing with structurally heterogeneous 3D-EM data sets are expected to make a significant contribution to our understanding of a wide range of molecular machines.

Acknowledgments

I am grateful to Tony Crowther for critically reading the chapter and to Pawel Penczek, Richard Hall, and Bjoern Sander for providing materials for Figs. 2, 3, and 5, respectively.

References

Agirrezabala, X., Lei, J., Brunelle, J. L., Ortiz-Meoz, R. F., Green, R., Frank, J. (2008). Visualization of the hybrid state of tRNA binding promoted by spontaneous ratcheting of the ribosome. *Mol. Cell* **32**, 190–197.

Alberts, B. (1998). The cell as a collection of protein machines: preparing the next generation of molecular biologists. *Cell* **92**, 291–294.

Baxter, W. T., Grassucci, R. A., Gao, H., Frank, J. (2009). Determination of signal-to-noise ratios and spectral SNRs in cryo-EM low-dose imaging of molecules. *J. Struct. Biol.* **166**, 126–132.

Berriman, J., Unwin, N. (1994). Analysis of transient structures by cryo-microscopy combined with rapid mixing of spray droplets. *Ultramicroscopy* **56**, 241–252.

Brink, J., Ludtke, S. J., Kong, Y., Wakil, S. J., Ma, J., Chiu, W. (2004). Experimental verification of conformational variation of human fatty acid synthase as predicted by normal mode analysis. *Structure* **12**, 185–191.

Brunger, A. T. (1992). Free R value: a novel statistical quantity for assessing the accuracy of crystal structures. *Nature* **355**, 472–475.

Cheng, K., Ivanova, N., Scheres, S. H., Pavlov, M. Y., Carazo, J. M., Hebert, H., et al. (2010). tmRNA.SmpB complex mimics native aminoacyl-tRNAs in the A site of stalled ribosomes. *J. Struct. Biol.* **169**, 342–348.

Cornish, P. V., Ermolenko, D. N., Noller, H. F., Ha, T. (2008). Spontaneous intersubunit rotation in single ribosomes. *Mol. Cell* **30**, 578–588.

Cuellar, J., Martin-Benito, J., Scheres, S. H., Sousa, R., Moro, F., Lopez-Vinas, E., et al. (2008). The structure of CCT-Hsc70 NBD suggests a mechanism for Hsp70 delivery of substrates to the chaperonin. *Nat. Struct. Mol. Biol.* **15**, 858–864.

Dube, P., Tavares, P., Lurz, R., van Heel, M. (1993). The portal protein of bacteriophage SPP1: a DNA pump with 13-fold symmetry. *EMBO J.* **12**, 1303–1309.

Elad, N., Clare, D. K., Saibil, H. R., Orlova, E. V. (2008). Detection and separation of heterogeneity in molecular complexes by statistical analysis of their two-dimensional projections. *J. Struct. Biol.* **162**, 108–120.

Elad, N., Farr, G. W., Clare, D. K., Orlova, E. V., Horwich, A. L., Saibil, H. R. (2007). Topologies of a substrate protein bound to the chaperonin GroEL. *Mol. Cell* **26**, 415–426.

Faruqi, A. R. (2009). Potential impact of silicon pixel detectors on structural biology. *Nucl. Instrum. Methods Phys. Res. A* **607**, 7–12.

Fischer, N., Konevega, A. L., Wintermeyer, W., Rodnina, M. V., Stark, H. (2010). Ribosome dynamics and tRNA movement by time-resolved electron cryomicroscopy. *Nature* **466**, 329–333.

Frank, J. (2006). Three-dimensional Electron Microscopy of Macromolecular Assemblies. Oxford University Press, New York.

Fu, J., Gao, H., Frank, J. (2007). Unsupervised classification of single particles by cluster tracking in multi-dimensional space. *J. Struct. Biol.* **157**, 226–239.

Gao, H., Valle, M., Ehrenberg, M., Frank, J. (2004). Dynamics of EF-G interaction with the ribosome explored by classification of a heterogeneous cryo-EM dataset. *J. Struct. Biol.* **147**, 283–290.

Gipson, P., Mills, D. J., Wouts, R., Grininger, M., Vonck, J., Kuhlbrandt, W. (2010). Direct structural insight into the substrate-shuttling mechanism of yeast fatty acid synthase by electron cryomicroscopy. *Proc. Natl. Acad. Sci. USA* **107**, 9164–9169.

Hall, R. J., Siridechadilok, B., Nogales, E. (2007). Cross-correlation of common lines: a novel approach for single-particle reconstruction of a structure containing a flexible domain. *J. Struct. Biol.* **159**, 474–482.

Herman, G. T., Kalinowski, M. (2008). Classification of heterogeneous electron microscopic projections into homogeneous subsets. *Ultramicroscopy* **108**, 327–338.

Heymann, J. B., Cheng, N., Newcomb, W. W., Trus, B. L., Brown, J. C., Steven, A. C. (2003). Dynamics of herpes simplex virus capsid maturation visualized by time-lapse cryo-electron microscopy. *Nat. Struct. Biol.* **10**, 334–341.

Julian, P., Konevega, A. L., Scheres, S. H., Lazaro, M., Gil, D., Wintermeyer, W., et al. (2008). Structure of ratcheted ribosomes with tRNAs in hybrid states. *Proc. Natl. Acad. Sci. USA* **105**, 16924–16927.

Kastner, B., Fischer, N., Golas, M. M., Sander, B., Dube, P., Boehringer, D., et al. (2008). GraFix: sample preparation for single-particle electron cryomicroscopy. *Nat. Methods* **5**, 53–55.

Klinge, S., Nunez-Ramirez, R., Llorca, O., Pellegrini, L. (2009). 3D architecture of DNA Pol alpha reveals the functional core of multi-subunit replicative polymerases. *EMBO J.* **28**, 1978–1987.

Kostek, S. A., Grob, P., De Carlo, S., Lipscomb, J. S., Garczarek, F., Nogales, E. (2006). Molecular architecture and conformational flexibility of human RNA polymerase II. *Structure* **14**, 1691–1700.

Kuhn, C. D., Geiger, S. R., Baumli, S., Gartmann, M., Gerber, J., Jennebach, S., et al. (2007). Functional architecture of RNA polymerase I. *Cell* **131**, 1260–1272.

Leschziner, A. E., Nogales, E. (2006). The orthogonal tilt reconstruction method: an approach to generating single-class volumes with no missing cone for ab initio reconstruction of asymmetric particles. *J. Struct. Biol.* **153**, 284–299.

Leschziner, A. E., Saha, A., Wittmeyer, J., Zhang, Y., Bustamante, C., Cairns, B. R., et al. (2007). Conformational flexibility in the chromatin remodeler RSC observed by electron microscopy and the orthogonal tilt reconstruction method. *Proc. Natl. Acad. Sci. USA* **104**, 4913–4918.

McMullan, G., Clark, A. T., Turchetta, R., Faruqi, A. R. (2009). Enhanced imaging in low dose electron microscopy using electron counting. *Ultramicroscopy* **109**, 1411–1416.

Moffat, K., Henderson, R. (1995). Freeze trapping of reaction intermediates. *Curr. Opin. Struct. Biol.* **5**, 656–663.

Nagayama, K., Danev, R. (2008). Phase contrast electron microscopy: development of thin-film phase plates and biological applications. *Philos. Trans. R. Soc. Lond. B Biol. Sci.* **363**, 2153–2162.

Nickell, S., Beck, F., Korinek, A., Mihalache, O., Baumeister, W., Plitzko, J. M. (2007). Automated cryoelectron microscopy of ''single particles'' applied to the 26S proteasome. *FEBS Lett.* **581**, 2751–2756.

Nickell, S., Beck, F., Scheres, S. H., Korinek, A., Forster, F., Lasker, K., et al. (2009). Insights into the molecular architecture of the 26S proteasome. *Proc. Natl. Acad. Sci. USA* **106**, 11943–11947.

Penczek, P. A., Frank, J., Spahn, C. M. (2006). A method of focused classification, based on the bootstrap 3D variance analysis, and its application to EF-G-dependent translocation. *J. Struct. Biol.* **154**, 184–194.

Penczek, P. A., Yang, C., Frank, J., Spahn, C. M. (2006). Estimation of variance in single-particle reconstruction using the bootstrap technique. *J. Struct. Biol.* **154**, 168–183.

Radermacher, M. (1988). Three-dimensional reconstruction of single particles from random and nonrandom tilt series. *J. Electron Microsc. Tech.* **9**, 359–394.

Radermacher, M. (2009). Chapter 1 Visualizing functional flexibility by three-dimensional electron microscopy reconstructing complex I of the mitochondrial respiratory chain. *Methods Enzymol.* **456**, 3–27.

Radermacher, M., Wagenknecht, T., Verschoor, A., Frank, J. (1986). A new 3-D reconstruction scheme applied to the 50S ribosomal subunit of E. coli. *J. Microsc.* **141**, RP1–RP2.

Rose, H. H. (2009). Future trends in aberration-corrected electron microscopy. *Philos. Transact. A Math. Phys. Eng. Sci.* **367**, 3809–3823.

Rosenthal, P. B., Henderson, R. (2003). Optimal determination of particle orientation, absolute hand, and contrast loss in single-particle electron cryomicroscopy. *J. Mol. Biol.* **333**, 721–745.

Sander, B., Golas, M. M., Luhrmann, R., Stark, H. (2010). An approach for de novo structure determination of dynamic molecular assemblies by electron cryomicroscopy. *Structure* **18**, 667–676.

Sander, B., Golas, M. M., Makarov, E. M., Brahms, H., Kastner, B., Luhrmann, R., et al. (2006). Organization of core spliceosomal components U5 snRNA loop I and U4/U6 Di-snRNP within U4/U6.U5 Tri-snRNP as revealed by electron cryomicroscopy. *Mol. Cell* **24**, 267–278.

Scheres, S. H., Carazo, J. M. (2009). Introducing robustness to maximum-likelihood refinement of electron-microscopy data. *Acta Crystallogr. D Biol. Crystallogr.* **65**, 672–678.

Scheres, S. H. W., Gao, H., Valle, M., Herman, G. T., Eggermont, P. P., Frank, J., et al. (2007). Disentangling conformational states of macromolecules in 3D-EM through likelihood optimization. *Nat. Methods* **4**, 27–29.

Scheres, S. H., Melero, R., Valle, M., Carazo, J. M. (2009). Averaging of electron subtomograms and random conical tilt reconstructions through likelihood optimization. *Structure* **17**, 1563–1572.

Scheres, S. H., Nunez-Ramirez, R., Gomez-Llorente, Y., San Martin, C., Eggermont, P. P., Carazo, J. M. (2007). Modeling experimental image formation for likelihood-based classification of electron microscopy data. *Structure* **15**, 1167–1177.

Scheres, S. H., Valle, M., Carazo, J. M. (2005). Fast maximum-likelihood refinement of electron microscopy images. *Bioinformatics* **21** (Suppl. 2), ii243–244.

Scheres, S. H., Valle, M., Grob, P., Nogales, E., Carazo, J. M. (2009). Maximum likelihood refinement of electron microscopy data with normalization errors. *J. Struct. Biol.* **166**, 234–240.

Scheres, S. H. W., Valle, M., Núñez, R., Sorzano, C. O. S., Marabini, R., Herman, G. T., et al. (2005). Maximum-likelihood multi-reference refinement for electron microscopy images. *J. Mol. Biol.* **348**, 139–149.

Shaikh, T. R., Barnard, D., Meng, X., Wagenknecht, T. (2009). Implementation of a flash-photolysis system for time-resolved cryo-electron microscopy. *J. Struct. Biol.* **165**, 184–189.

Shatsky, M., Hall, R. J., Nogales, E., Malik, J., Brenner, S. E. (2010). Automated multi-model reconstruction from single-particle electron microscopy data. *J. Struct. Biol.* **170**, 98–108.

Sigworth, F. J. (1998). A maximum-likelihood approach to single-particle image refinement. *J. Struct. Biol.* **122**, 328–339.

Sigworth, F. J., Doerschuk, P. C., Carazo, J. M., Scheres, S. H. (2010). An introduction to maximum likelihood methods in cryo-EM. *Methods Enzymol.* **482**, 263–294.

Stagg, S. M., Lander, G. C., Pulokas, J., Fellmann, D., Cheng, A., Quispe, J. D., et al. (2006). Automated cryoEM data acquisition and analysis of 284742 particles of GroEL. *J. Struct. Biol.* **155**, 470–481.

Suloway, C., Pulokas, J., Fellmann, D., Cheng, A., Guerra, F., Quispe, J., et al. (2005). Automated molecular microscopy: the new Leginon system. *J. Struct. Biol.* **151**, 41–60.

Suloway, C., Shi, J., Cheng, A., Pulokas, J., Carragher, B., Potter, C. S., et al. (2009). Fully automated, sequential tilt-series acquisition with Leginon. *J. Struct. Biol.* **167**, 11–18.

Tagare, H. D., Barthel, A., Sigworth, F. J. (2010). An adaptive Expectation-Maximization algorithm with GPU implementation for electron cryomicroscopy. *J. Struct. Biol.* **171**, 256–265.

Tagare, H. D., Sigworth, F., Barthel, A. (2008). Fast, adaptive expectation-maximization alignment for Cryo-EM. *Med. Image Comput. Comput. Assist. Interv.* **11**, 855–862.

Tama, F., Brooks, C. L., III (2002). The mechanism and pathway of pH induced swelling in cowpea chlorotic mottle virus. *J. Mol. Biol.* **318**, 733–747.

Tama, F., Wriggers, W., Brooks, C. L., III, (2002). Exploring global distortions of biological macromolecules and assemblies from low-resolution structural information and elastic network theory. *J. Mol. Biol.* **321**, 297–305.

Valle, M., Sengupta, J., Swami, N. K., Grassucci, R. A., Burkhardt, N., Nierhaus, K. H., et al. (2002). Cryo-EM reveals an active role for aminoacyl-tRNA in the accommodation process. *EMBO J.* **21**, 3557–3567.

van Heel, M. (1984). Multivariate statistical classification of noisy images (randomly oriented biological macromolecules). *Ultramicroscopy* **13**, 165–183.

van Heel, M., Frank, J. (1981). Use of multivariate statistics in analysing the images of biological macromolecules. *Ultramicroscopy* **6**, 187–194.

van Heel, M., Stoffler-Meilicke, M. (1985). Characteristic views of E. coli and B. stearothermophilus 30S ribosomal subunits in the electron microscope. *EMBO J.* **4**, 2389–2395.

Voss, N. R., Yoshioka, C. K., Radermacher, M., Potter, C. S., Carragher, B. (2009). DoG Picker and TiltPicker: software tools to facilitate particle selection in single particle electron microscopy. *J. Struct. Biol.* **166**, 205–213.

Wang, H. W., Noland, C., Siridechadilok, B., Taylor, D. W., Ma, E., Felderer, K., et al. (2009). Structural insights into RNA processing by the human RISC-loading complex. *Nat. Struct. Mol. Biol.* **16**, 1148–1153.

Weis, F., Bron, P., Rolland, J. P., Thomas, D., Felden, B., Gillet, R. (2010). Accommodation of tmRNA-SmpB into stalled ribosomes: a cryo-EM study. *RNA* **16**, 299–306.

White, H. E., Orlova, E. V., Chen, S., Wang, L., Ignatiou, A., Gowen, B., et al. (2006). Multiple distinct assemblies reveal conformational flexibility in the small heat shock protein Hsp26. *Structure* **14**, 1197–1204.

White, H. E., Saibil, H. R., Ignatiou, A., Orlova, E. V. (2004). Recognition and separation of single particles with size variation by statistical analysis of their images. *J. Mol. Biol.* **336**, 453–460.

White, H. D., Thirumurugan, K., Walker, M. L., Trinick, J. (2003). A second generation apparatus for time-resolved electron cryo-microscopy using stepper motors and electrospray. *J. Struct. Biol.* **144**, 246–252.

Wong, S. S., Wong, L. J. (1992). Chemical crosslinking and the stabilization of proteins and enzymes. *Enzyme Microb. Technol.* **14**, 866–874.

Yoshioka, C., Pulokas, J., Fellmann, D., Potter, C. S., Milligan, R. A., Carragher, B. (2007). Automation of random conical tilt and orthogonal tilt data collection using feature-based correlation. *J. Struct. Biol.* **159**, 335–346.

Yu, L., Snapp, R. R., Ruiz, T., Radermacher, M. (2010). Probabilistic principal component analysis with expectation maximization (PPCA-EM) facilitates volume classification and estimates the missing data. *J. Struct. Biol.* **171**, 18–30.

Zhang, W., Kimmel, M., Spahn, C. M., Penczek, P. A. (2008). Heterogeneity of large macromolecular complexes revealed by 3D cryo-EM variance analysis. *Structure* **16**, 1770–1776.

AUTHOR INDEX

A

B

C

D

E

F

G

H

I

J

K

L

Q

R

S

Z

SUBJECT INDEX

Note: The letters '*f*' and '*t*' following the locators refer to figures and tables respectively.

A

B

C

D

E

O

P

R

S

T

U

V

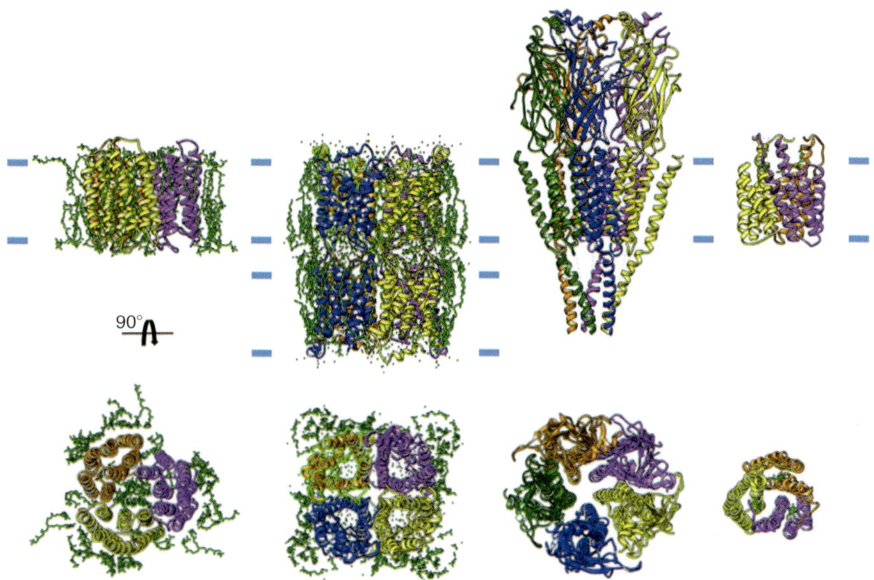

IBAN UBARRETXENA-BELANDIA AND DAVID L. STOKES, CHAPTER 2, FIG. 2. Selection of high-resolution membrane protein structures solved by electron crystallography. From left to right: a trimer of bacteriorhodopsin, a double-layer of the tetrameric AQP0, the hetero-pentameric acetylcholine receptor, and a trimer of gluthathione transferase 1. The approximate boundaries of the bilayer are indicated by short blue lines, and individual lipid molecules present in the structure are shown in green.

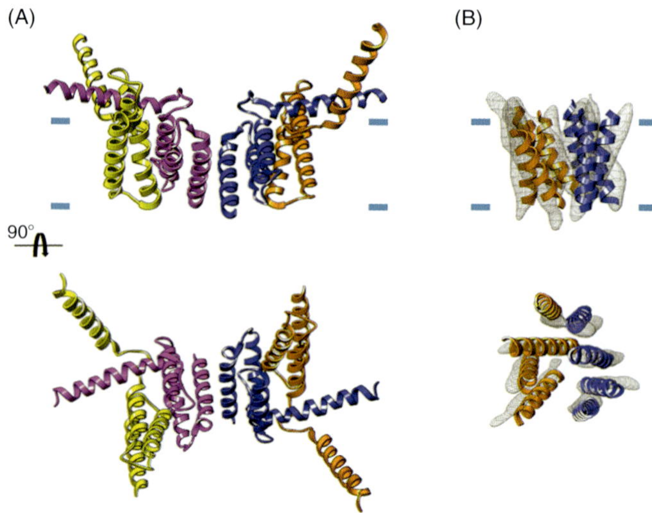

Iban Ubarretxena-Belandia and David L. Stokes, Chapter 2, Fig. 3. Membrane crystals preserve the native structure of membrane proteins. (A) The original X-ray structure showing a dimer of the multidrug-resistance antiporter from *E. coli* EmrE solved from 3D crystals formed with detergent-solubilized protein (pdb 1S7B). (B) The electron crystallographic map of monomeric EmrE (emdb emd-1087) solved at a resolution of 7 Å fitted with the model derived by Fleishman et al. (2006) (pdb 2I68). In the X-ray structure, two of the helices are protruding radically from the presumed bilayer plane, which is indicated by the short blue lines.

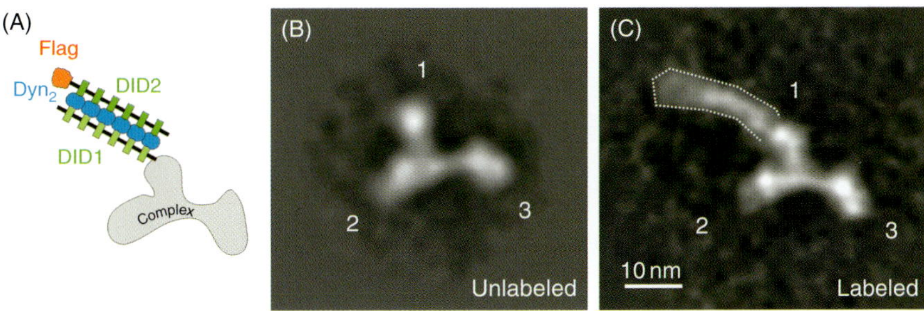

Bettina Böttcher and Katharina Hipp, Chapter 3, Fig. 2. Molecular pointer—DID1-Dyn2-DID2 (adapted from Flemming et al., 2010). (A) Labeling scheme: the site of interest in the complex (gray) is tagged with a dynein light chain-interacting domain from Pac11 (DID1, green). This domain binds to dynein light chain dimers (Dyn2, blue) which form globular domains and to a second dynein light chain-interacting domain from Nup 159 (DID2, green) which is coupled with a Flag tag (red). The Flag tag enables purification of the labeled complex. The fully assembled label resembles pearls on a string. (B) The unlabeled Nup-84 complex consists of Nup120-Nup145C-Nup85-Seh1-Sec13 and forms three prominent domains (indicated by numbers). (C) Seh1 of the Nup84 complex is labeled with the DID1-Dyn2-DID2 label, which forms a prominent molecular pointer (outlined by a white, dotted line).

Bettina Böttcher and Katharina Hipp, Chapter 3, Fig. 3. Pseudo-atomic model of a complex (Diepholz et al., 2008). For many of the subunits and subcomplexes of the V-ATPase, high-resolution structural models exist or can be generated by homology modeling. Placing these models (ribbon diagrams) into the 3D-map of the whole complex (gray) generates a pseudo-atomic model for most of the V-ATPase. For some subunits, no high-resolution structural information exists. Their position and shape can be deduced by comparison of the density map with the pseudo-atomic model (e.g., subunit D, solid purple density; subunit a, solid blue density).